Chemicals and Materials from Renewable Resources

ACS SYMPOSIUM SERIES **784**

Chemicals and Materials from Renewable Resources

Joseph J. Bozell, EDITOR
National Renewable Energy Laboratory

American Chemical Society, Washington, DC

Library of Congress Cataloging-in-Publication Data

Cheemicals and materials from Renewable Resources / Joseph J. Bozell, editor.

p. cm.—(ACS symposium series ; 784)

Includes bibliographical references and index.

ISBN 0–8412–3727–1

1. Chemicals—Congresses. 2. Renewable natural resources—Congresses.

I. Bozell, Joseph J., 1953- II. Series.

TP201 .CS18 2001
660—dc21 00–50220

The paper used in this publication meets the minimum requirements of American National Standard for Information Sciences—Permanence of Paper for Printed Library Materials, ANSI Z39.48–1984.

Distributed by Oxford University Press

PRINTED IN THE UNITED STATES OF AMERICA

Foreword

The ACS Symposium Series was first published in 1974 to provide a mechanism for publishing symposia quickly in book form. The purpose of the series is to publish timely, comprehensive books developed from ACS sponsored symposia based on current scientific research. Occasionally, books are developed from symposia sponsored by other organizations when the topic is of keen interest to the chemistry audience.

Before agreeing to publish a book, the proposed table of contents is reviewed for appropriate and comprehensive coverage and for interest to the audience. Some papers may be excluded to better focus the book; others may be added to provide comprehensiveness. When appropriate, overview or introductory chapters are added. Drafts of chapters are peer-reviewed prior to final acceptance or rejection, and manuscripts are prepared in camera-ready format.

As a rule, only original research papers and original review papers are included in the volumes. Verbatim reproductions of previously published papers are not accepted.

ACS Books Department

Contents

Preface

A symposium, *Chemicals and Materials from Renewable Resources,* was held at the 218th National Meeting of the American Chemical Society in New Orleans, Louisiana, from August 22 to August 26, 1999. This book contains a collection of some of the papers presented during the 1.5 days of the symposium.

Each new oil crisis seems to bring a rediscovery of the renewable to chemicals field. Yet the work in this area remains very diverse, making it difficult to present a coherent front to those who might fund new efforts in basic and applied research. In organizing the symposium, we were faced with a similar problem: Do we focus on a single narrow topic or present a broader sample of several aspects of the field? We eventually chose a middle road on a basis of an idea, which is described more fully in Chapter 1, that our understanding of how one can selectively manipulate renewable feedstocks requires (1) a broader range of renewables-based building blocks and (2) a much greater understanding of how transformations of those building blocks can be controlled at a molecular level. It is our hope that this sampling of different research efforts will stimulate a broader interest in this field and will promote better communication among those groups working in related areas.

At this writing, the price of crude oil is again on the rise and the American public is facing sharply higher gasoline prices. The United States possesses a vast amount of renewable and sustainable feedstocks that could supply a significant proportion of our chemical needs while not sacrificing our supply of necessary food, feed, and fiber requirements. Perhaps these chapters will serve to catalyze a much broader research effort in this fascinating field.

JOSEPH J. BOZELL
National Renewable Energy Laboratory
1617 Cole Boulevard
Golden, CO 80401

Chapter 1

Chemicals and Materials from Renewable Resources

Joseph J. Bozell

National Renewable Energy Laboratory, 1617 Cole Boulevard, Golden, CO 80401

Abstract: A symposium entitled "Chemicals and Materials from Renewable Resources" was held as part of the ACS National Meeting in August, 1999. This introductory chapter attempts to show that a good case can be made for the use of renewables as chemical feedstocks. While economic considerations certainly play a role, greater use of renewables is primarily prevented by a lack of technology development. The breadth of technology available for high yield, selective manipulation of renewable raw materials pales in comparison to that available in the petrochemical industry. It is hoped that the papers collected in this monograph will introduce the reader to new technology available for the use of renewables, and serve as a catalyst for broader coordination of efforts in this developing field.

The Case for Renewables

The symposium on which this book is based is certainly not the first to describe the concept of using renewables as chemical feedstocks. Well into the 20th century, renewable feedstocks supplied a significant portion of the nation's chemical needs. The chemurgy movement of the 1930s, led by such notables as William Hale and Henry Ford, promoted the use of farm products as a source of chemicals, with the belief that "*anything that can be made from a hydrocarbon could be made from a carbohydrate" (1)*. It is only in the period of time between 1920 and 1950 that we have witnessed the transition to a nonrenewables based economy *(2)*.

A vast amount of renewable carbon is produced in the biosphere; about 77×10^9 tons is fixed annually, an amount that could supply almost all domestic organic chemical needs, currently about 7 - 8% of our total nonrenewables consumption *(3)*. When measured in energy terms, the amount of carbon synthesized is equivalent to about ten times the world consumption *(4)*. Cellulose, the most abundant organic chemical on earth, has an annual production of about 100×10^9 tons. Lignin production by the pulp and paper industry is 30 - 50×10^6 tons/year *(5)*. Yet our chemical feedstock supply is overwhelmingly dominated by nonrenewable carbon. Some of the largest industrial users of renewable feedstocks are the pulp and paper industry and the corn wet milling industry. However, the pulp and paper industry devotes only a small part of its production to chemicals, while the corn wet milling industry is focused largely on starch and its commercial derivatives, ethanol and corn syrup. Given such a plentiful potential source of chemicals and products, what are the barriers that stand between the promise of renewables and their wider use by the chemical industry?

Several advantages are frequently associated with renewables:

- The use of biomass has been suggested as a way to mitigate the buildup of greenhouse CO_2 in the atmosphere *(6)*. Since biomass uses CO_2 for growth through photosynthesis, the use of biomass as a feedstock results in no net increase in atmospheric CO_2 content when the products break down in the environment *(7)*.
- It is generally acknowledged that increased use of biomass would extend the lifetime of the available crude oil supplies. The Royal Dutch/Shell group has developed several scenarios for the impact of biomass on chemicals and fuel production. One scenario projects a biomass market in the first half of the 21st century of $150 billion/year, with up to 30% of worldwide chemical and fuel needs supplied by renewable feedstocks in the same time period *(8,9)*.
- A chemical industry incorporating a significant percentage of renewable materials is more secure because the feedstock supplies are domestic, leading to a lessened dependence on international "hot spots". The 1991 and 1996 events between Iraq and the U. S. and the immediate response of the oil

futures market is a graphic example of the hazards associated with dependence on politically unstable regions for crude oil supplies.

- Biomass is a more flexible feedstock than is crude oil. Crude oil is formed and its composition set by geological forces. The diversity of building blocks from biomass offers a great opportunity for the production of a range of chemicals as wide as that available from nonrenewables. With the advent of genetic engineering, the tailoring of certain plants to produce high levels of specific chemicals is also possible.
- Building blocks isolated from crude oil are not oxygenated, yet many of the final products of the chemical industry are. Biomass derived materials are often highly oxygenated. There are few general ways to add oxygen to hydrocarbons, and many of them require the use of toxic reagents (chromium, lead, etc) in stoichiometric amounts resulting in severe waste disposal problems *(10)*.

Moreover, increased use of renewable feedstocks could address other, broader issues:

Feedstock Needs Globally - Recent work has attempted to model when world oil production will peak *(11)*. This work, based on methodology developed in the 1950s for predicting U. S. oil production, projects that global production will begin to decline sometime in the next 5 - 10 years. The earlier work accurately predicted the decline in domestic oil production that occurred in the mid 1970s. Demand will not decrease in line with production. U.S. energy consumption has increased by more than 28 percent (about 21 quadrillion btu) during the last 25 years, but more than half of the overall energy growth of the last 25 years (about 11 quadrillion btu) has occurred during the last 6 years *(12)*. Other feedstock sources will be needed.

Domestic Energy Consumption - The U. S. annually consumes about 94 quads of energy *(12,13)*. Of this 94 quads, 35 are used by industry, and almost 8 quads are used in the production of chemicals and paper. This is a significant energy target, and one that could be addressed by a greater use of renewables.

These potential benefits would seem to indicate that renewables hold promise as a feedstock complementary to those used by the chemical industry. *Why isn't renewable carbon in its many forms (cellulose, lignin, monomeric and oligomeric carbohydrates, oils, terpenes, etc.) more widely used as a chemical feedstock?*

An answer to this question must begin with an economic comparison. Certainly the primary driver for greater domestic use of renewable feedstocks is cost because it will be the chemical industry that needs to be convinced of the value of renewables. That argument is only won by favorable economics. As shown in several studies, not all renewables-based materials are reasonable targets for biomass based processes, primarily for economic reasons *(14)*.

However, a number of polymeric (cellulose, lignin) and monomeric (carbohydrates: glucose, xylose; other materials: levulinic acid) materials are available at production costs that make them comparable to many nonrenewable raw materials *(15)*. Strong cost competition would exist from the simpler nonrenewable building blocks; it is difficult to beat ethylene at \$0.16/lb. However, many examples exist of low cost, readily available renewable building blocks that could be incorporated into the domestic feedstock supply stream. A cost argument is not sufficient to render renewables unattractive as feedstocks. Table I summarizes some sales or production costs of a few typical renewable building blocks (depending on the source, the costs are current as of 1995-1999).

Table I - Costs of Some Selected Renewable Feedstocks

Material	\$/kg	\$/lb	Cost Type	Source
Polymers				
Cellulose	0.44 - 1.10	0.20 - 0.50	Production	3
Lignin	0.07 - 0.13	0.03 - 0.06	Production	Fuel value
Carbohydrates				
glucose	0.60 - 1.10	0.27 - 0.50	Sales	1
	0.13 – 0.26	0.06 – 0.12	Production	3
xylose/arabinose	0.07 – 0.13	0.03 – 0.06	Production	3
sucrose	0.40	0.18	Sales	2
lactose	0.65	0.30	Sales	2
	0.50 – 1.50	0.23 – 0.68	Sales	1
fructose	0.90	0.41	Sales	1
sorbitol	1.60	0.73	Sales	1
Other				
Levulinic acid	0.18 – 0.26	0.08 – 0.12	Production	3

1. M. Bols, "Carbohydrate Building Blocks", Wiley-Interscience, New York (1996)
2. F. Lichtenthaler and S. Mondel, *Pure Appl. Chem.*, **1997**, *69*, 1853.
3. Range of estimates from discussions with various industrial sources.

One must also examine possible barriers in the overall production of chemicals. However, there are some interesting similarities between the renewables and petrochemical industries. Both face the same three general issues for producing chemicals. First is an issue of supply. Both industries need to know the source of their feedstock, its lifetime, and methods for its removal from the structures in which it is found. Second, both industries must face issues of separation by determining what components are present in the feedstock and how these components are separated from one another. Finally, both industries face the issue of conversion. Once the building blocks are removed from the feedstock, they must be converted to products. Of these three areas, issues of supply and separation are generally understood for both renewables and petrochemical refining.

Technology Development Issues

Thus, we are left to examine the differences in issues of conversion. What stands between the concept and realization of a vibrant renewables to chemicals industry is *technology development*. The world understands and can manipulate existing nonrenewable feedstocks. In contrast, the level of technology development for similar manipulation of renewables lags far behind. Where nonrenewables particularly excel in comparison to renewables is that understanding of nonrenewables is much greater at a *molecular* level. The nonrenewables industry has gained amazing control over the behavior of their primary building blocks. Efficient functionalization and rearrangement of their primary building blocks is a cornerstone of the industry. And importantly, these transformations are achieved selectively (i. e., without the formation of unwanted byproducts) and in very high yield (i. e., all of the starting feedstock is converted into the next product) *(16)*.

In contrast, the analogous use of renewables suffers from a much narrower range of discrete building blocks, fewer methods to convert those building blocks to other materials, a lack of fundamental understanding of how to convert starting raw materials (lignin, carbohydrates, oil crops, protein, biomass polymers, etc.) into single products in high yield, and a lack of information about the properties and performance available from those products. We are faced with the puzzle of possessing an almost limitless source of raw material in the United States, while being unable to effectively convert it to a wide range of useful products.

It was in this context that the symposium was organized. The hope was to attract participants that could speak to this technology gap, and describe new research that would address the shortcomings renewables currently face. What are the new building blocks available from renewables? What properties do they exhibit? How does one selectively transform biomass polymers into single products? What is known about the mechanisms of these transformations? What new methodologies hold promise for improving the utility of renewables? How can synthesis introduce desirable structural features into renewable building blocks? How can the structural features inherent to renewable building blocks most effectively be used?

As a result, a broad range of topics was covered at this symposium, reflecting the diversity that currently exists in this field. Yet the symposium could have been much broader: a truly comprehensive symposium on renewables could have included topics such as agricultural practices, new genetic engineering techniques for plants, biofuels, modern separation techniques, marketing, engineering of renewables-based processes, education, and a wide range of other related disciplines. But to keep the program manageable, the organizers decided to focus as closely as possible on new building blocks and materials that can be derived from renewables. We were particularly interested in papers that improved understanding of renewables at the *molecular* level. When phenomena are defined and understood at a molecular level, control of processes is a result. Process control and understanding allow for the effective design of molecular structure for function, for example, the production of a new polymeric material, or new commodity chemical with new properties.

Overview of Papers

In the following chapters the reader will find: 1) overview material; 2) novel monomeric building blocks; 3) biochemical processes; and 4) new polymeric materials. Even with a program limited to these four topics, the organizers received some comments of concern regarding the content of this symposium, since it seemed to lack a clear focus. This is certainly a danger. However, at this stage in the development of the renewables industry, we felt the intent should be to provide a sampling of several areas, with focus on specific topics of interest to be used in future symposia.

Part 1 - Overview material

The first paper in the book is included as an important contrast to any unwarranted renewables fervor. The contribution by Gerngross describes the need for careful life cycle analyses and shows how renewables should not automatically be assumed to be good, just because they are "green". The production of PHAs in comparison to polystyrene is used as an example. The following chapter by McLaren presents a general vision for renewables, based on the Vision 2020 process being used by the U. S. Department of Energy.

Part 2 - New chemicals and intermediates. Chemical processes and mechanism.

This section describes several new building blocks available from renewables, and how those building blocks can be converted by chemical processes into other materials. The contributions by Moens and Olson describe new uses for levulinic acid, an important renewable building block available inexpensively from cellulosic wastes *(17)*. Kiely's paper on carbohydrate diacids describes approaches to selective glucose oxidation, the preparation of new polymers made from oxidized glucose, and where these materials might find use in the chemical industry. Witczak overviews uses of levoglucosenone, an interesting renewable building block available from starch pyrolysis. The paper describes some of the important structural details that control its reactivity, and presents the chemical details and the utility of new technology. The paper by Gandini reports the author's work in converting furans, available from renewable feedstocks, into new polymeric materials. Gravitis discusses how furfural and levoglucosan can be prepared and isolated from biomass as novel renewable building blocks using new tandem catalysis technology. The contribution from Brown describes recent work in biomass pretreatment technology to improve the production of levoglucosan, a building block that is on the verge of becoming a large scale fine chemical intermediate.

Part 3 - Bioprocesses/Biotechnology

The use of biotechnology to convert fermentable sugars into chemicals is certainly one of the most rapidly growing areas of the chemistry of renewables. Frost's paper describes work carried out on understanding and tailoring components of the shikimic acid cascade, and how this work will lead to a wide range of chemicals from one of the key intermediates, dihydroshikimate. The contribution from Baker describes approaches to and properties of new polylactides, of particular current importance with the planned commercialization of a polylactic acid process by Cargill Dow Polymers. Finally, Nghiem's paper presents details on the development of a bioprocess for producing succinic acid, and how it has gone from a laboratory scale program to planned commercial development.

Part 4 - New Materials from Renewables

Two papers point to the use of renewables in the production of interesting polymeric materials based on renewables. The contribution from Kelley describes a model compound study on the reactivity of pyrolysis oils, and how this information can be used in the formulation of new renewables based adhesives. Glasser describes the use of steam explosion technology to facilitate the simultaneous production of lignin and cellulose esters, useful in the production of thermoplastic structural polymers from wood.

Conclusions

It is hoped that the reader will view this collection of diverse topics as a starting point to stimulate greater interest in the field. Obviously, a key component for fully developing this field will be the willingness of funding agencies to support such work. As pointed out by Lichtenthaler:

"The prospects are promising, if one is prepared to generously invest into basic research...research can be planned, results cannot (!). Thus, the only other thing then needed is patience - patience until the "fruits" from renewables, in upgraded, value added form, can be harvested."*(18)*.

As workers in the field, we hope someone is listening.

Acknowledgements

I would like to thank both the Carbohydrate and Cellulose, Paper and Textile divisions of the American Chemical Society for cosponsoring what I hope will be the first in a series of such symposia. I would also like to extend my deepest appreciation to the authors that took part of their valuable time to contribute to this effort.

References

1. Interestingly, the chemurgical movement was not initiated as a result of a shortage of other feedstocks, but rather as a way to use farm surpluses: Clark, J. P., "Chemurgy". In *Kirk-Othmer Encyclopedia of Chemical Technology,* 4th ed., Vol. 5. Kroschwitz, J. I., Howe-Grant, M., Eds., John Wiley & Sons, New York, 1993; p. 902; Hale, W. J., *The Farm Chemurgic*, The Stratford Co., Boston, MA, 1934.
2. For an overview of the history of this transition, see Morris, D.; Ahmed, I. *The Carbohydrate Economy: Making Chemicals and Industrial Materials from Plant Matter*, Institute for Local Self Reliance, Washington, D. C., 1992.
3. Donaldson, T. L.; Culberson, O. L. *Energy* **1984**, *9*, 693; Lipinsky, E. S. *Science* **1981**, *212*, 1465; Hanselmann, K. W. *Experientia* **1982**, *38*, 176.
4. Indergaard, M.; Johansson, A.; Crawford, Jr., B. *Chimia* **1989**, *43*, 230.
5. Kuhad, R. C.; Singh, A. *Crit. Rev. Biotech.* **1993**, *13*, 151.
6. Hall, D. O.; House, J. I. *Sol. Energy Mater. Sol. Cells* **1995**, *38*, 521.
7. This statement presumes that the CO_2 contained in standing trees, debris, and soil is not released and that consumed biomass is replaced by fresh plantings, i.e., the system must be sustainable. See Harmon, M. E.; Ferrell, W. K.; Franklin, J. F. *Science* **1990**, *247*, 699.
8. *The Evolution of the World's Energy System*, Group External Affairs, Shell International Limited, SIL Shell Centre, London, SE1 7NA; Romm, J. J., and Curtis, C. R. *Atlantic Monthly* **1996**, *277*, (4), 57.
9. The concept of "diminishing supplies" is subjective. During the oil crises of the 1970s, projections concluded that oil prices would be about $100/barrel by the late 20th century. Despite recent price increases, crude oil is still relatively inexpensive. A projection by the International Energy Authority predicts that the price of oil (in 1993 dollars) will rise only to $28/barrel by 2050.

10. Haines, A. H. *Methods for the Oxidation of Organic Compounds*, Academic Press, New York, 1985, 1988.
11. Anon., *Science* **1998**, *281*, 1128.
12. Energy Information Administration, Annual Energy Review 1997, DOE/EIA-0384(97). (Washington, DC, July 1998); information also available at http://www.eia.doe.gov.
13. 1 quad = 1 quadrillion (10^{15}) btu.
14. Bozell, J. J.; Landucci, R. eds., *Alternative Feedstocks Program Technical and Economic Assessment: Thermal/Chemical and Bioprocessing Components*, available from the author at the National Renewable Energy Laboratory, 1993.
15. Several sources of information are available for costs: Bols, M. *Carbohydrate Building Blocks*, Wiley-Interscience, New York, 1996; Lichtenthaler, F.; Mondel, S. *Pure Appl. Chem.* **1997**, *68*, 1853; Chemical Market Reporter, the National Corn Growers Association.
16. The amazing diversity of today's chemical industry is described in an excellent book: Weissermel, K.; Arpe, H.-J.; *Industrial Organic Chemistry*, 3rd edition, VCH, Weinheim, 1997.
17. Bozell, J. J.; Moens, L.; Elliott, D. C.; Wang, Y.; Neuenscwander, G. G.; Fitzpatrick, S. W.; Bilski, R. J.; Jarnefeld, J. L.; *Resour. Conserv. Recycl.* **2000**, *28*, 227.
18. Lichtenthaler, F. W.; *Carb. Res.* **1998**, *313*, 69.

Chapter 2

Polyhydroxyalkanoates: The Answer to Sustainable Polymer Production

Tillman U. Gerngross

Thayer School of Engineering, Dartmouth College, Hanover, NH 03755 (email: tillman.gerngross@dartmouth.edu)

PHAs have frequently been championed as a solution to sustainable polymer production. This is because they can be produced from renewable raw materials and are biodegradable upon disposal. However, an inventory of materials and energy required to produce these polymers reveals a rather discouraging picture. In most categories quantifying environmental impact (land use, resource depletion and emissions to air and water) PHA production by fermentation scores worse than conventional petrochemical polymer production.

The question of, "How can we supply energy and materials to a society that has grown increasingly dependent on non-renewable, fossil resources?", will be at the center of a vigorous scientific and political debate in the years to come. With oil, gas and coal reserves expected to be depleted in 75, 66 and 666 years respectively *(1,2)*, the search for alternative energy and feedstocks has already become the focus of concerted research efforts and is expected to intensify as we enter the 21st century. Dartmouth has arguably taken a leadership role in this area by dedicating its entire chemical engineering program to the discovery of biological alternatives to conventional commodity production of fuels and chemicals. However when considering options to make chemicals and materials from renewable resources one has to remember what provided the impetus for developing such strategies in the first place. Two fundamental objectives come to mind: (a) reducing the depletion of finite resources and (b) reducing the release of harmful pollutants to air, water and land during manufacture. While the transition to a renewable resource based economy must provide benefits in both, (a) reduced resource depletion and (b) emissions reduction, we currently do not have sufficient information to assess the potential environmental impact of such a transition. Our case study on making PHAs from a renewable resource has shown that these two objectives can be diametrically opposed. We have found that, in the case of PHA production by fermentation, the former objective (achieving lower emissions) has been displaced by a mainly rhetorical commitment to the latter (sustainability or reduced resource depletion) *(3,5)*. This is because it was generally assumed that the goal of sustainable resource utilization is inherently compatible with environmental preservation. It is this fundamental tenet that requires further understanding and has become a focus of our research interests. Promoting the conversion to technologies based on renewable raw materials has to be based on thorough understanding of both the processes that are to be substituted and the processes that we intend to substitute with. Accordingly, we sought to analyze one of the options currently under consideration for making polymers from renewable resources and found a picture substantially more ambiguous than originally anticipated. Our findings are specific to the production of PHAs from corn in a fermentation process and should not be generalized to include other approaches to produce biopolymers from renewable resources. The analysis of those processes is still ongoing and will be the subject of future review.

PHAs as an Alternative to Petrochemical Polymers

Amongst the various alternative polymers developed to date, PHAs, a class of aliphatic microbial polyesters, have been viewed as one of the leading candidates to replace conventional plastics on a large industrial scale *(4)*. Like their petrochemical counterparts, PHAs are moldable, water insoluble, thermoplastic polymers. The most common PHA, poly-3-hydroxybutyrate (PHB), is a stiff, high-melting aliphatic polyester similar to many industrial polyolefins. Unlike polyolefins, however, PHAs can be synthesized by microorganisms, which can produce and store the polymer in the form of intracellular inclusions at levels exceeding 80% of the cell dry mass (see Figure 1). These microbial polymers can be made entirely from glucose in a fermentation process, and, in addition to offering favorable material properties, PHAs are completely biodegradable. Thus, replacing conventional plastics with these biopolymers has been promoted as a desirable approach to solid waste management and sustainable polymer production *(5,6)*.

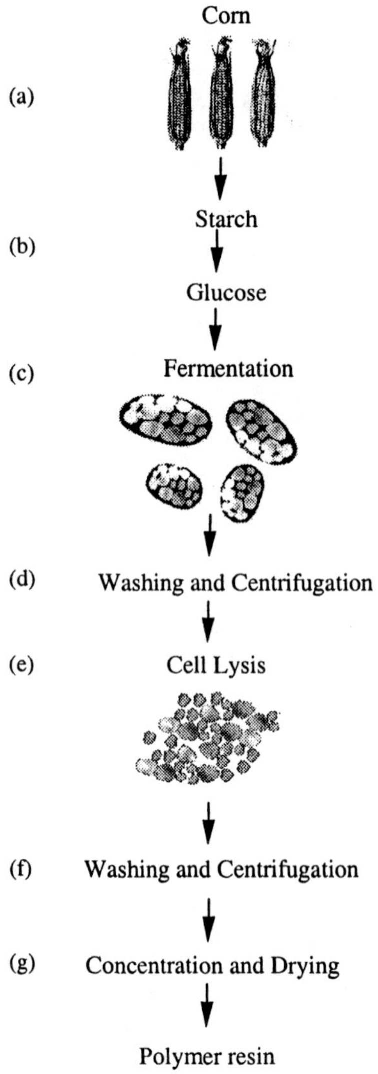

Figure 1. The PHA process from corn to polymer.

The analyzed PHA production process consists of the following steps: (a) growing and harvesting corn (b) processing said corn to yield glucose (c) sterilizing the medium and conducting a fermentation process by which glucose is converted to PHA, (note cells containing PHA polymer in the form of intracellular inclusions) (d) recovering the biomass containing the polymer and washing it by centrifugation, (e) disrupting the cell wall and releasing the polymer from the cells, (f) washing the polymer by centrifugation and (g) drying the polymer slurry to a powder which can be processed to a final consumer product. (Reproduced with permission from reference 3. Copyright 1999.)

The Benchmark of Sustainability

Several factors contribute to the environmental impact and the degree of sustainability of a given product or material. In many instances, however, impact is not caused by the product *per se,* but by the consumption of raw materials and the release of waste products during manufacture. Hence a "cradle to grave" analysis, which incorporates manufacturing practices, energy input-output, and overall material flows, becomes the benchmark for assessing environmental impact and sustainability *(7,8)*. In this paper we make an effort to analyze the sustainability of two polymer processes by comparing the overall energy and raw material requirements to produce the biological polymer PHA, and the fossil fuel based polymer polystyrene (PS).

Contrary to the widespread belief that PHAs are a sustainable alternative to polymer production, we find a surprisingly high latent energy content associated with their fermentative production. If we include the utilities to make glucose from corn, producing PHA in a fermentation process would consume 22% more steam, 19 times the amount of electricity and more than 7 times the amount of water compared to a PS process. We conclude therefore, that fermentative PHA production despite the use of renewable agricultural feedstocks, consumes significantly more energy, releases more net greenhouse gases *(9)* and therefore is no more sustainable than conventional petrochemical polymer production.

The Energy Dilemma

While only PS is directly derived from fossil oil, both polymers require energy in their manufacture, which in the U.S. is largely generated by combusting fossil resources *(10)*, hence both polymers have a latent fossil fuel content. In order to determine the environmental benefits of substituting a petrochemical process with a biological process, we compiled an energy and feedstock inventory of a "theoretical" large-scale PHA process and compared it with a conventional polystyrene process.

Oil refineries, through the fractionation of crude oil and natural gas, produce various primary refinery products. The vast majority of these products are used as transportation fuels and for heating. However, a minor fraction (less than 4%) is processed into petrochemical intermediates that serve as feedstocks for the polymer industry. One such intermediate is styrene, which can be polymerized to yield PS, a polymer that finds wide application in the fast food and packaging industry. Analysis of PS production revealed an energy requirement, which is equivalent to the consumption of 1.26kg fossil oil for each kg of polystyrene *(8)*. In other words one needs 2.26kg of fossil oil to generate energy (1.26kg) and feedstocks (1kg) to produce 1kg of PS (see Table I).

Table I. Direct raw material requirements for the production of polystyrene (PS) versus polyhydroxyalkanoate (PHA)

	Per kg polymer	
Item	*PS*	*PHA*
Glucose (kg)	none	3.3[a]
Petroleum fractions (kg)	1.78[b]	none
Inorganic salts (g)	20[b]	149[c]
Water (L)	4[b]	26[d]

[a] The yield of PHA on glucose in bacterial fermentations is around 30%.

[b] Petroleum fractions serve as a feedstock and as an energy source in the manufacture of PS. The production of 1kg of PS requires a total of 2.26kg of fossil fuels, of which 0.48kg can be directly assigned to the production of steam and electricity. Of the remaining 1.78kg, which serve as feedstocks, a fraction is recycled and also used as an energy source. For details see *(8)*.

[c] Inorganic salts are required for the fermentation process in the form of potassium and sodium phosphate, ammonia and phosphoric acid as well as trace elements *(11)*. However in the subsequent energy analysis we only account for the energy to produce ammonia which is about 109 kg per ton polymer – this because ammonia production accounts for most of the energy required in the form of inorganic salts.

[d] Water is required to fill the fermenter and for four washes. It does not include cooling water, which is expected to be recycled through a chiller or a cooling tower.

SOURCE: Reproduced with permission from reference 3. Copyright 1999.

Polyhydroxyalkanoate production on the other hand is based on corn, which is one of the more energy intensive agricultural crops. Corn production accounts for about 44%, 44% and 55% of total fertilizer, insecticide and herbicide use in the U.S., respectively *(12)*. Several researchers have attempted to quantify the energy required to till, irrigate, fertilize and harvest corn and have found the process to consume between 2226 and 6722KJ per kg of corn grain *(12,13,14)*. By incorporating primary energy usage patterns of U.S. farms this translates into the combustion of between 50 and 160kg of fossil fuels per ton of corn grain *(15)*.

Making fermentable glucose from corn involves growing and harvesting corn, transferring it to a wet mill where it is fractionated and processed to yield gluten meal, oil, starch and sugars, most importantly glucose and fructose. In 1991 the average energy to process 1kg of corn grain required about 4375KJ (see Table II), which equates the combustion of 130kg of fossil fuels per ton processed grain *(16)*.

Table II. Energy required to produce raw materials for the production of PHA

	Per kg of raw material		
Item	*Glucose*	*Inorganic Salts*	*Water*
	8129KJ[a]	30,135KJ[b]	5.6 KJ[c]

[a] Energy to produce 1.52 kg of corn grain (60% starch) and process the grain to yield 1kg of glucose minus energy for co-products (33% of corn milling energy). Calculation: 1.52 kg x 2442KJ/kg (14) + 4417KJ/kg (corn wet milling). The entire corn wet milling industry consumed 147 trillion KJ of energy by combusting externally purchased fossil fuels or buying external electrical power *(16)*. In the same year a total of 33.66 million tons of corn grain were processed *(17)*. Hence the average energy input can be estimated to be about 4375 KJ per kg processed corn grain and therefore the production of 1 kg of dextrose from 1.52 kg of corn grain (starch content 60%), consumes about 6650 KJ *(18)*. Of that energy 33% are allocated to the production of co-products (corn meal etc.) leaving 4417KJ per kg of glucose. In addition we contacted several engineering firms, active in the construction of corn wet mills (Nofsinger, Kansas City, MO and Process Systems Inc., Memphis, TN) to confirm these estimates. We obtained values ranging from 3718 - 5631KJ for the production of one kg of glucose from corn grain.

[b] Energy to produce one kg of ammonia, which is used in the fermentation to control pH and provide nitrogen *(19)*.

[c] Energy to filter and pump water to point of use *(20)*.

SOURCE: Reproduced with permission from reference 3. Copyright 1999.

A Best Case Scenario for Fermentative PHA Production from Corn

Several fermentation-based PHA processes have been described and could serve as a basis for this analysis *(21,22,23)*. While they may differ in the choice of microorganism and other process characteristics, all reported PHA fermentations share the following features: (a) very high PHA concentrations can be obtained, commonly above 100g/l, (b) biological polymer synthesis is an exothermic process and therefore requires external cooling, (c) 2.4kg of carbon dioxide are emitted per kg of polymer produced during the fermentation, (d) all processes are aerobic and therefore require considerable aeration and agitation, (e) cell wall disruption is required to release the polymer from the microorganism, (f) between 0.15 and 0.3 kg of biological waste are generated per kg of polymer (g) and approximately 3.33 kg of glucose are required to produce 1kg of PHA. We will base our analysis on an optimistic high cell density PHA fermentation as outlined in Figure 1. The fermentation process is assumed to last 48 hours and produce an impressive 190g/l biomass of which 150g/l are PHA (79% polymer content). After fermentation, we assume the recovery of 100% of the polymer produced in the fermentation. Our analysis shows that 2.39kg of fossil resources (gas, oil and coal) are required to produce 1kg of PHA, thereby exceeding the amount required to produce an equal amount of PS (Tables III and IV).

While the consumption of fossil resources does not differ greatly between the two processes the emissions reveal a discouraging fact. PHA production requires the combustion of the entire 2.39kg for energy production whereas PS production combusts only 1.26kg of the 2.26 kg fossil fuel required for its manufacture, the balance being used as a feedstock. Our energy consumption estimates have been very conservative and are far below those of other researchers that have assigned energy requirements to PHA fermentation processes *(22)*. In fact van Wegen and coworkers, who analyzed the fermentation process alone, assigned energy values exceeding ours by 57% for electricity and 467% for steam. The net effect would be fairly drastic resulting in an overall fossil fuel consumption of 3.73kg per kg of PHA.

More Energy Equates More Emissions

While fermentative PHA production offers a route by which a renewable resource can be converted into a thermoplastic polymer, the energy requirements to perform such a process are troublesome. Notwithstanding this dilemma, the "closed carbon cycle" of PHA production has often been cited as evidence for its "sustainability" (Figure 2, A). This assertion was made without providing any data or accounting for either material or energy inputs and outputs, which are required to draw such a conclusion. A more comprehensive and accurate picture is presented in Figure 2, B where not only the flux of carbon raw materials but also the flux of materials required for the production of process energy, namely fossil fuel equivalents (FFE), are accounted for. Since the production of PHA requires the combustion of more fossil fuels than PS, more greenhouse gas emissions are to be expected.

Two forms of carbon emission have to be considered in this context: (a) temporary carbon, which is carbon that comes from the corn plant and (b) fossil carbon, which is combusted to produce energy. Any emissions generated from temporary carbon, may it be in the fermentation process through microbial activity, the biodegradation of the corn stalks on the field, or degradation of the polymer upon disposal, are not considered to be net carbon emissions. An equal amount of carbon will be sequestered in the plant in the following crop cycle and therefore this type of carbon is only temporarily released into the atmosphere. The combustion of FFE on the other hand leads to a net carbon emission, that would otherwise not occur and as such has to be weighed against carbon emissions generated in a conventional polymer process. Since energy consumption is significantly higher in the PHA case, it reasons that carbon dioxide emissions are higher as well. In fact carbon dioxide emissions associated with the production of PHA are more than 60% higher then those for PS production. Naturally, other emissions associated with power generation from fossil resources such as SO_2, NO_x and N_2O are expected to follow a similar trend. In light of Article 3 of the Kyoto protocol, which specifically calls for the reduction of carbon dioxide and N_2O, an increase of these emissions stands in direct opposition to universally supported efforts aimed at improving air quality and reducing global warming.

Conclusions

In light of these findings we conclude that replacing conventional polymers with fermentation derived PHAs does not appear to be a useful approach if a sustainable production of polymers is the desired outcome. Other benefits such as the biodegradability and biocompatibility of PHAs could justify the expense of considerable fossil resources; however, those benefits would have to be quantified and evaluated separately. As such, the usefulness of biodegradability itself has been put into question. The disposal of paper has shown that the rate of degradation in landfills is not only slow *(26)*, but results in greenhouse gas emissions (methane and carbon dioxide) as well as leachates with increased biological oxygen demand. Proper incineration, on the other hand, would not only produce less harmful greenhouse gases *(7)* but more importantly allows for the partial recovery of energy expended during manufacture.

By using corn as a feedstock and employing current fermentation technology, the environmental benefits of producing PHAs are questionable even under rather favorable assumptions. While the biological production from renewable resources certainly has the potential to conserve fossil resources, the case study submitted in this paper demonstrates that such an approach can also have the reverse effect. Therefore future assessments of biological processes have to incorporate not only the use of raw materials, which are mostly renewable, but also address the indirect consumption of non-renewable energy sources required to carry out such a process.

Table III. Energy required for the fermentative production of PHA from glucose

Item	*Electricity (kWh)*	*Steam (kg)*	*Fossil Fuel Equivalent[a] (kg)*
		Per kg PHA	
Fermentation			
Media sterilization[b]	none	0.45	0.02
Agitation[c]	0.32	none	0.09
Aeration[c]	1.27	none	0.35
Cooling[d]	0.76	none	0.21
Downstream			
Centrifugation and Washing[e]	0.50	none	0.14
High pressure Homogenizer[f]	1.97	none	0.54
Centrifugation and Washing[e]	0.50	none	0.14
Evaporation[g]	none	0.33	0.02
Spray drying[h]	none	2.00	0.10
Total	5.32	2.78	1.59

[a] Amount of fossil fuel (kg) consumed to generate the electricity and steam listed in the same row. Conversion: Electrical energy (kWh) x 0.272 (kWh/kg) = FFE (kg). In 1997 the U.S. average for producing 1 KWh of electrical power, from all power sources (including geothermal, hydroelectric, nuclear and alternative power generation) required the direct combustion of 0.272 kg of fossil resources *(10)*. These resources were 83% coal, 13% natural gas and 3.5% petroleum. To convert steam (kg) to FFE (kg): Steam (kg) x Heat of evaporation of water (2400KJ/kg) / 47,219 kJ/kg (Heat of combustion of natural gas) = FFE (kg). Calculation assumes 100% efficiency, the use of natural gas for steam generation and no heat loss.
[b] Medium is continuously sterilized to 143 °C for 30seconds; 68% of energy is recaptured through a heat exchanger and used to pre-warm the incoming medium.
[c] Agitation and aeration for this type of aerobic fermentation with very high cell densities is estimated to require 5W/l of power input. Power is delivered by mechanical agitation and compressed air and amounts to 1.59kWh per kg of PHA. *(24)*
[d] Cooling: We are considering the use of a fully jacketed 114,000 liter fermenter (3:1, H:D) that has additional cooling coils on the inside providing a total cooling area 266 m^2

or 2.32 m^2 per m^3. Under the assumed production schedule (48h and 190g/l of biomass formed) we have to remove approximately 17.6 W per liter or a total of 2 million J/s *(25)*. Coolant is provided through cooling towers and chillers which deliver about 48,700KJ and 18,000KJ of cooling per kWh input respectively *(20)*

[e] Centrifugation: See reference *(22)* We reduced the energy input by 66% to reflect the energy savings from our substantially higher solids content.

[f] Cell disruption: See reference *(22)*

[g] Evaporation is performed with a triple effect evaporator. A pre-concentrated slurry containing 30% solids is concentrated to 50% solids. This step requires the evaporation of 1.33kg of water and requires 0.33kg of steam per kg PHA. Further concentration is not feasible because of the high viscosity of the slurry.

[h] The final slurry is processed to a powder employing a spray dryer. Dry air is required to provide process heat and evaporate the remaining water. The energy required is generally about double the amount of the water that has to be evaporated from the slurry. Hence about 2kg of steam is required to remove 1kg of water leaving us with 1kg of dry powder.

SOURCE: Reproduced with permission from reference 3. Copyright 1999.

Table IV. Energy and fossil fuel equivalents (FFE) required in the production of PS versus PHA

	Per kg polymer			
Item	*PS*		*PHA*	
	Energy	FFE (kg) [a]	Energy	FFE (kg) [a]
Production of raw materials	see below [b]	1.78 [b]	31,218KJ[c]	0.80
Utilities				
Steam [d]	7.0 kg	0.40	2.78 kg	0.14
Electricity [e]	0.30 kWh	0.08	5.324 kWh	1.45
Total		2.26		2.39

[a] Fossil fuels required to produce the energy and raw material in the corresponding energy column taking into account the primary fuel usage patterns for each industry. Example: In the corn wet milling energy is generated from natural gas (37%), coal (48%), and petroleum (10%) *(14)*. Therefore, for each MJ of energy, currently consumed by the corn wet milling industry 370,000KJ are generated by combusting 7.84 kg of natural gas etc. Petroleum, coal and gas have a heat of combustion of 42,000KJ/kg, 25,788KJ/kg and 47,218KJ/kg, respectively. Other conversions see Table III.

[b] Feedstock required to produce 1kg of PS from crude oil. The feedstock for direct polymer synthesis is included as well as a fraction that is lost in the process and recycled for energy generation. Since this type of process does not allow for a clear allocation of feedstocks and energy numbers include both; feedstock and process energy. It does not however include the itemized steam and electricity energy listed below. See also *(8)*.

[c] Energy (including steam and electrical power) to produce glucose, ammonia and water. Calculations: 3.33 (Yield of PHA on glucose) x 8,128KJ (see Table II) + 4,028 KJ (Energy to produce 0.109kg of ammonia) + 120KJ (Energy to provide 26l of process water). See also Table I.

[d] The production of one kg of steam requires the combustion of 0.058kg of residual fuel oil during PS production. PHA fermentation most likely will use natural gas for steam generation and therefor only requires 0.0508kg of gas to produce the same amount of steam.

[e] For conversion of kWh to FFE see Table III.

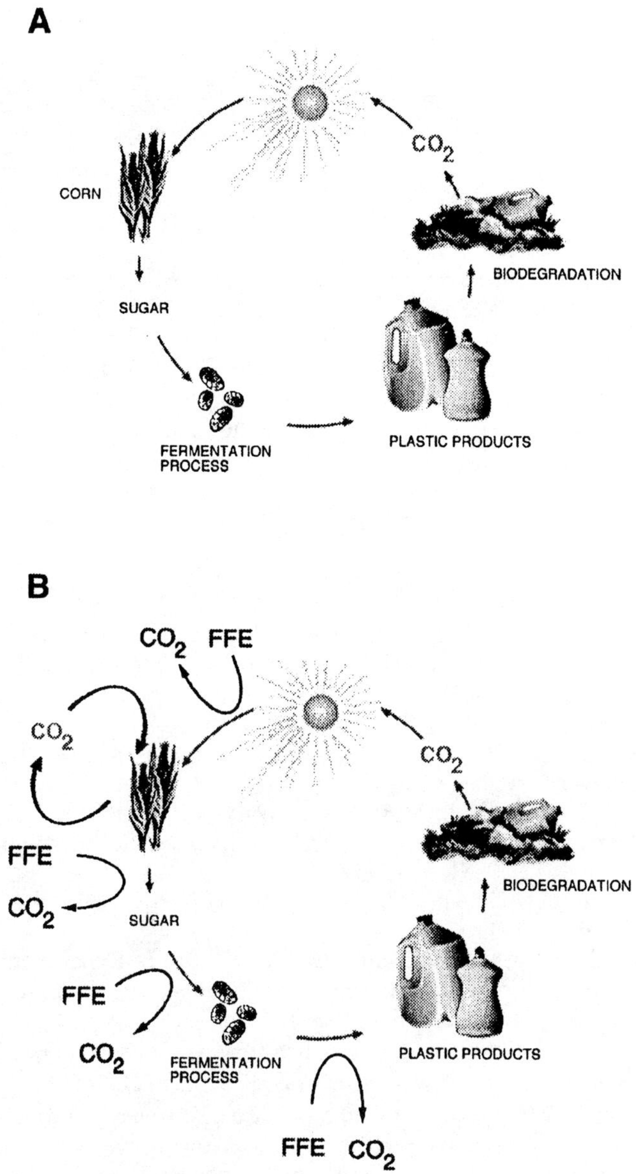

Figure 2. Carbon cycle for the production of PHA from corn

(A) Carbon cycle of raw materials for PHA production from corn.
(B) Carbon cycle of PHA production including fossil fuel requirements to grow corn and provide process energy for PHA production. Temporary and fossil carbon emissions are shown in gray and black respectively.

References

1. World Energy Council, 1995, http://www.worldenergy.org.
2. Campbell, C. J.; Laherrere, J. H. *Scientific American* **1998**, *March,* 78-83.
3. Gerngross, T. U. *Nat. Biotech.* **1999**, *17,* 541-544.
4. Pool, R. *Science* **1989**, *245,* 1187-1189.
5. Williams, S. F.; Peoples, O. P. *Chemtech* **1996**, *26,* 38-44.
6. Lee, S. Y. *Nature-Biotechnology* **1997**, 17-18.
7. Hocking, M. B. *Science* **1991**, *251,* 504-505.
8. Hocking, M. B. *Environmental Management* **1991**, *15,* 731-747.
9. While all the carbon dioxide that is emitted during the fermentation is indeed of plant origin and therefore should not be viewed as a net green house gas contribution, this is not the case for the 2.5 kg of fossil fuels that have to be combusted to provide process energy.
10. Energy Information Administration, Department of Energy, http://www.eia.doe.gov/.
11. Byrom, D. US Patent 5,364,778 (Zeneca Limited, London, England, United States, 1994).
12. Smil, V.; Nachman, P. T.;Long, V. I. *Energy Analysis and Agriculture, An Application to U.S. Corn Production* (Westview Press, Inc., Boulder, Colorado, 1983).
13. Pimentel, D.; Pimentel, S. *New Directions for Agriculture and Agricultural Research* Dahlberg, K. A. Ed. (Rowman and Allanheld, Totawa, New Jersey, 1986).
14. Shapouri, H.; Duffield, J. A.; Graboski, M. S. *Estimating the Net Energy Balance of Corn Ethanol*, Agricultural Economic Report Number 721 (USDA, 1995).
15. Fossil fuels used in U.S. farming are: coal 10%, natural gas 49%, and petroleum 40%.
16. Energy Information Administration of the U.S. Department of Energy, Manufacturing Consumption of Energy 1991.
17. Source: USDA-Economic Research Service.
18. The census of manufacturing data for energy consumption is based on calendar years, whereas corn processing data is based on the harvesting cycle starting with September, 1 of a given year. The data correlated are for the calendar year 1991 and the harvesting cycle starting September 1, 1990 and ending August 31, 1991.
19. Hocking, M. B. *Handbook of Chemical Technology and Pollution Control* (Academic Press, San Diego, CA, 1993).
20. Ulrich, G. D. *A Guide to Chemical Engineering Process Design and Economics* (John Wiley & Sons, New York, 1984).
21. Lee , S. Y.; Choi, J. I. *Polym. Degrad. Stabil.* **1998**, *59,* 387-393.

22. van Wegen, R.J; Ling, Y.; Middelberg, A. P. J. *Trans IChemE, Part A* **1998**, *76,* 417-426.
23. Lee, S. Y. *TIBTech* **1996**, *14,* 431-438.
24. Charles, M.; Wilson, J. *Bioprocess Engineering.* B. J. Lydersen, N. A. D'Elia, K. L. Nelson, Eds. (1994).
25. Atkinson, B.; Mavituna, F. *Biochemical Engineering and Biotechnology Handbook* (Stockton Press, New York, NY, ed. 2nd, 1991).
26. Reese, K. M. *Chem. Eng. News* **1989**, *67,* 60.

Chapter 3

The Vision for Renewable Resources

James S. McLaren

Inverizon International Inc., 2214 Stoneridge Terrace Court, Chesterfield, MO 63017

Introduction

For the past thousand years or so, humans have demonstrated a remarkable ability to discover new scientific principles and to develop technology in ways that typically benefit society as a whole. Not only have we avoided the Malthusian prediction of impending disaster, we have actually supported an ever-growing population at an ever-increasing standard of living. As we look forward we can foresee continued change, but predicting future outcomes remains a challenging task. Will the beginning of the 21st century prove Malthus correct, will science come to the rescue yet again, or will technology be the downfall of society as is being predicted by some activist ideologies?

The U.S. "Vision for Renewable Resources" is concerned with the future supply of basic inputs, and materials, that end up in the thousands of products found in modern society, such as solvents, plastics, pesticides, structural materials, pharmaceuticals, and fibers (*1*). The Vision does not attempt to be predictive of a particular future outcome. Rather it takes a more pragmatic approach of reviewing where we are today, exploring what future needs are likely to be, then setting challenging goals that would maintain current societal standards on a sustainable basis. The expectation is that the technology required to meet those goals will arise from encouraging and supporting the appropriate types of scientific enquiry. The directional priorities and areas that require integrative approaches are outlined in the roadmap entitled "Plant/Crop-Based Renewable Resources 2020" (*2*).

Irrespective of national boundaries or cultural beliefs, most people agree that "SUSTAINABLE WELL-BEING" (S_{wb}) is a reasonable condition to strive for. However, problems may arise due to different ideological definitions of this term. The definition used here implies a temporal factor but does not include going backwards in time to some romantic notion of the past. Rather, it implies that sustainable means going forward with repeatable improvements. Thus, such a state can be described as in 1:

$$S_{wb} = (R_{pn} + R_r) \times (C_s + C_t + C_i) \quad (1)$$

S_{wb} = sustainable well-being
R_{pn} = resources that are pre-existing natural
R_r = resources that are renewable in a short time frame, e.g., annual
C_s = capability due to scientific advance level
C_t = capability due to technology application
C_i = capability due to investment in infrastructure

Capability (C) requires the appropriate level of scientific knowledge, plus methods being available to implement such knowledge, plus investment in utilizing this advance in a beneficial way. A simple analogy is that in a world of candles we can make many shapes and sizes, and invest in making incremental improvements in each. However, no matter how fancy the candles may be, it is not the same as having the scientific advance called electricity. Resources (R) can be viewed as natural, where some substance is taken as pre-existing in nature (R_{pn}) and can be collected or harvested. Most natural systems were sufficient as sources of inputs until the early to mid-1900's. Driven by the pressure of population demand, the intensification of "extraction" has often resulted in concomitant problems within the local ecosystem itself. It is a false contention that decreased material consumption and reliance on R_{pn} alone can suffice for future needs, yet we cannot continue to harvest the pre-existing environment with indefinite abandon. The dilemma is to recognize the human right to exist, and to have reasonably equal opportunity to utilize global resources, but to achieve this in a manner that is sustainable.

Sustainable denotes a positive term and means repeatable over time, therefore, it is closely related to the term renewable (R_r). Sustainable does not necessarily relate to maintaining any particular pre-existing environmental situation (e.g. one that just happens to be current). While sustainable could include the status quo in certain time-scales, it also does imply that "change" is consistent with collecting repeat outputs. Thus, resources are taken as being both pre-existing "natural" plus those that can be renewed on a frequent basis in the human time-scale ($R_{pn} + R_r$).

Hydrocarbons PLUS Renewables

At one time in history, the majority of materials used for most products came from pre-existing natural sources or from agricultural production. This period was followed by the petrochemical revolution that improved living standards through innumerable different ways. Some groups believe that somehow "natural" materials are better than "man-made" materials, but most of those people hold this opinion from the comfort of a dwelling insulated with blown fiberglass, heated by advanced fuels, stocked by ample food supplies, and disinfected by powerful cleaning fluids.

There is no question that hydrocarbon-based products are beneficial and they have become ubiquitous in society. However, it would be very short-sighted to expect continued reliance on a finite pre-existing fossil fuel source as the sole platform for inputs. Therefore, the Vision for renewable resources proposes a future where material needs are met by a combination of both hydrocarbons and other renewable inputs from agriculture/forestry (Figure 1).

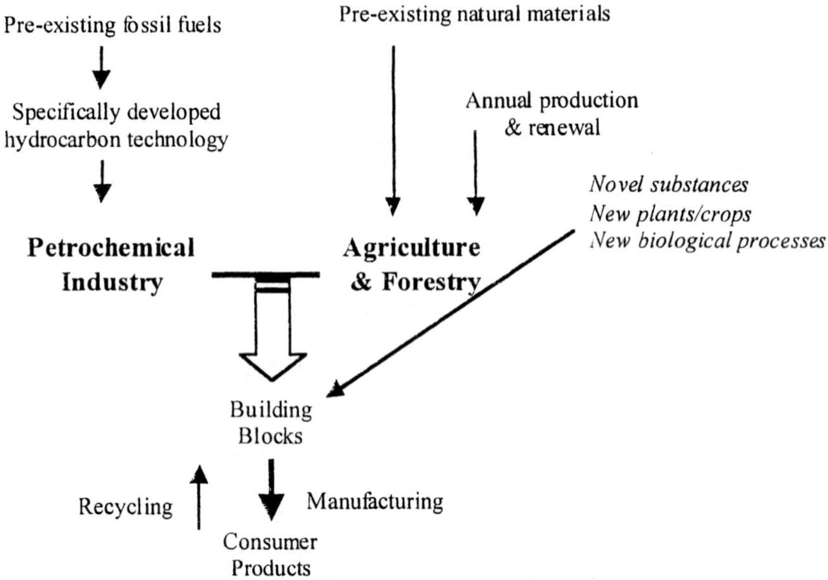

Figure 1. Representation of the expected complimentary nature of the petrochemical and agricultural industries in providing building blocks. The relative quantities supplied from each area will continue to change with time.

At the present point in time, hydrocarbon-based supply makes up about 95% of industrial inputs, with only about 3-5% arising from agriculture/forestry (excluding the use of wood for furniture and house building). Such a large reliance on the fossil fuel base is problematical due to:

a) the fact that fossil fuels are a finite source and non-renewable within human time-frames. While debate exists as to the supply volume and duration, any crisis will likely be driven more by dramatic price volatility and this will occur long before the supply runs out.
b) concerns over political stability and national security (*3*).
c) increasing anthropogenic impacts on environmental parameters through continued high levels of fossil fuel use, for example, global temperature elevations associated with higher atmospheric concentration of certain "greenhouse" gases (*4*).

Global demand for chemicals, materials, and products will continue to increase as a function of population growth plus elevated per capita use. This latter aspect will be especially relevant as the developing countries continue to emulate the more industrially developed nations in productivity (Figure 2).

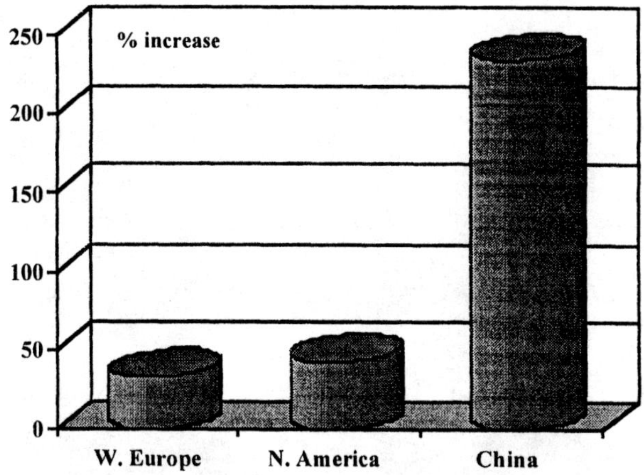

Figure 2. Projections of the increase in chemical output for 2010 as a percent of that recorded in 1996. Adapted from U.K. Chemical Industries Association (5).

In addition to the success story of over the past 50 years, there will be considerable further development of technology within the petrochemical industry: for example, the top five U.S. chemical companies are projecting R&D spending at $3.59 billion for the year 2000 (*6*). However, even with the improvements in efficiency that come from the application of new technology, the consumptive use-rate will be large enough to generate a net demand increase over current levels. Therefore, there is a high probability that incremental supply will be required to meet future demand projections.

The need for more material inputs and higher volumes of building blocks could be satisfied by increasing the use of fossil fuels. However, this may not be the best option due to the potential problems with high fossil fuel use mentioned above. Moreover, the U.S. already imports almost 60% of crude oil needs (*7*) and alternative sources of input materials are required for longer-term solutions. The Vision for renewable resources has been based on an assumption that we remain very reliant on hydrocarbons but do not want to increase the proportional use of fossil fuels. Therefore, to meet demand in the year 2020, the use of hydrocarbons from fossil fuels was kept about constant which created a "gap" between projected supply and projected demand. Figure 3 shows that the Vision calls for that gap to be filled through the utilization of renewable resources. Success would result in a 5-fold increase in renewable resources by 2020. While, under current paradigms, this is viewed as an aggressive goal, the total use of renewables would still remain below 25% of overall demand for building blocks and industrial inputs.

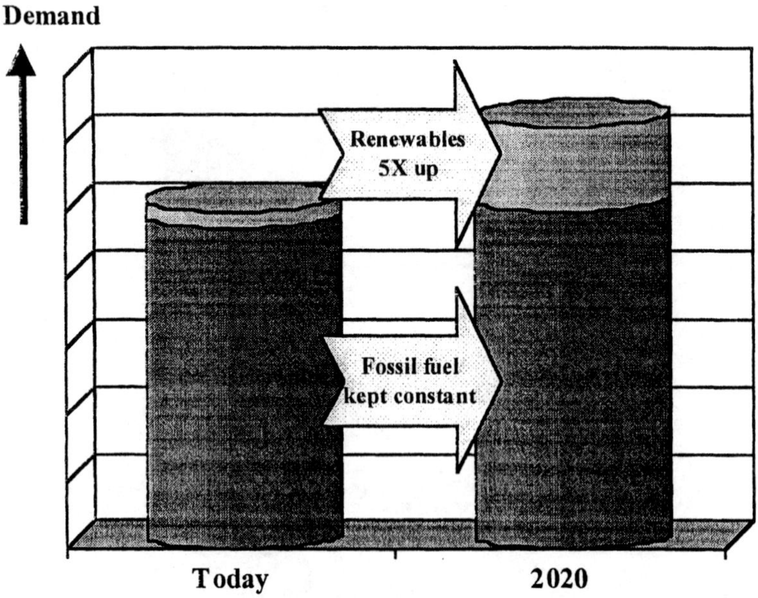

Figure 3. The contribution of material inputs and basic building blocks for manufacturing that comes from renewable resources and from fossil fuels, today and for the 2020 vision.

Achieving such a visionary situation will require significant new research in several areas related to renewable resources. If the knowledge were available today and we could calculate how to achieve this goal then it would be a simple matter of application of existing knowledge and would not require funding for additional research. However, the majority of experts involved have indicated that "new science" is required to meet the expectations for future contributions from renewable resources. Thus, inputs on research needs were collected from a broad range of stakeholder constituencies and were analyzed to form priority groupings within short-, medium- and long-term time-frames. The resulting document called the "Technology Roadmap for Plant/Crop-Based Renewable Resources 2020" provides directional input for the research required to meet the challenging goals (*2*).

Key Barriers

Identifying the major barriers that exist for enhanced utilization of renewable resources helped to set the priorities for research. The four main areas were considered to be:

Utilization

Utilization implies that functional performance is well understood and that input materials may be used *per se*, be blended, be altered in some way, or be used to make a novel product. Attempts to fit pre-existing materials into a hydrocarbon processing stream have not resulted in large volume use because the functionality of the materials is often out of alignment. The misalignment may be in technical attributes such as functional reactivity in a molding process, or may be in apparent economic disadvantage.

Economic comparisons are essential but must be made in perspective, and must go beyond a cost of goods mentality alone. At one time the cost of synthetic rubber was 20-fold the cost of natural rubber yet, today, the planet is not covered with plantations. Utilization efficiency, potential volume, potential product range characteristics, environmental impact costs, and sustainability must all be considered in economic calculations.

Altering the manufacturing process itself to accommodate different types of inputs may be required to allow the use of more renewable resources. Why would renewable materials fit a specifically designed hydrocarbon-based system? A related alternative is to explore making novel products that serve the same use need but are made from different materials: over time we have had drinking vessels made of various materials such as stone, fiber, glass, ceramics, and plastic. A major issue in utilization is that we must continue to rely heavily on hydrocarbon-based manufacturing, while making a gradual transition to alternative renewables. Transition economics tend to create a hurdle that inhibits this type of commercial progress. Some other driving force, such as support via a policy directive, is often required to overcome the initial economic barrier to increased utilization.

Processing

Over the past 50 years, the petrochemical industry has evolved into a sophisticated system for processing and distributing hydrocarbons. There has also been a huge investment in the infrastructure to support this particular source of carbon skeletons. Two major aspects that must be considered for the future processing efficiency of renewable resources are separations/conversions, and geographical location.

Location is important since most hydrocarbon processing facilities are not located in central positions relative to the large supply of agriculturally-derived renewable materials. Thus, raw material transport costs to existing facilities result in an inherent disadvantage for renewables. Currently, the best processing alignment for renewable type inputs is between corn production and wet milling operations. Over the next 20 years, we expect that a geographical shift will occur that co-locates more of the processing facilities with major supply areas, in the interior of the country.

A major barrier in the processing area is the lack of efficient technology for the separation and/or conversions of plant parts. For example, trees and plant stems have a high content of lignocellulose. Such a complex molecular matrix results in several problems in terms of processing efficiency. New approaches to thermochemical (*8*) or enzymatic "extraction" (*9*) are required to allow expanded use of the large amount of lignocellulosic material produced on an annual basis. Crop seeds contain carbohydrate, protein, oils, and hundreds of secondary chemicals (Figure 4) that require novel separations, especially those applicable to dilute aqueous streams. Currently, fermentation is a major route for the processing of such diverse constituents in the input materials (*10*). Future processing will require new advanced methods and may include more effective combinations of mechanical, thermochemical, and biological approaches.

Production

Currently, the U.S. produces a vast amount of renewable resources with major crops being grown on around 400 million acres, forestry being practiced on 650 million acres, and rangeland covering some 800 million acres. Much of this existing production goes into the feed/food supply (Figure 4) or is utilized in the paper industry, or in structural wood products. However, the associated lignocellulosic biomass that is "wasted" has been estimated at over 1200 billion pounds (10) which could provide a useful base of renewable inputs.

In addition to the current levels of production and co-produced material that is not well-utilized, there is a very large potential to be gained from advanced biological modification. Most crops and trees have been genetically modified over time through non-specific cross breeding and phenotypic selection. While very successful in improving production, those approaches are analogous to shooting in the dark when compared to highly specific genetic engineering methods that are just beginning to be used with considerable benefits (*11*). This type of approach will allow significant further improvements in productivity – enhancing both the volume per unit area and sustainable contribution because of more efficient use of inputs per unit of output.

In addition to enhanced unit productivity there may be some opportunities for expanded area contributions. While expanding highly productive land area is unlikely due to competitive anthropogenic uses, there could be more development of marginal land areas for renewable resources. For example, the development of crops that carry traits for improved tolerance to low moisture would allow more use of dryland areas. Moreover, neither the Great Lakes nor the many smaller lakes throughout the nation are used for any significant production today, but could be developed into large reservoirs of renewable resources.

Plant Science

Beyond plant breeding for higher yields and improved unit productivity, there is a significant opportunity to genetically enhance the composition of crops and trees. Although most modern cultivars have been genetically bred, we remain dependent on the "plant" constituents (Figure 4). For example, crop seeds still develop primarily for reproductive purposes and humans have adapted to process such seeds and grains. Recently, it has become apparent that constituents can be altered to more suit the requirements for utilization rather than for plant reproduction. In most cases in U.S. agriculture, crops grown for seed are distinct and produced in specific fields.

Genomics, the understanding of genes and how they function, has become a major breakthrough area of science – initially due to the Human Genome Project and the promise of genetic prevention of many diseases. Facilitated by the National Corn Genome Initiative, crops and trees have more recently become the subject of various expanded genome programs *(12)*, with the potential to allow precise manipulation of plant components and composition. Rather than having plants be a diverse mix of various chemical substances, it will be possible to shift metabolism towards certain desirable materials. Additionally, it will become much easier to search for, and identify, useful genes throughout the plant kingdom, and to transfer these into high yielding elite cultivars. Early genomics projects have focused on the feed, food and biomass production aspects of growing plants. The application of genomics to the production of particular materials within plants will make a large contribution to future renewable resources.

Current Examples

There are several areas where considerable research and development has been initiated to expand inputs from renewable resources. The aspects being addressed range from new technology being applied to existing materials, to new approaches for the generation of novel materials. In some cases, considerable capital has already been invested in "biorefinery" facilities, while

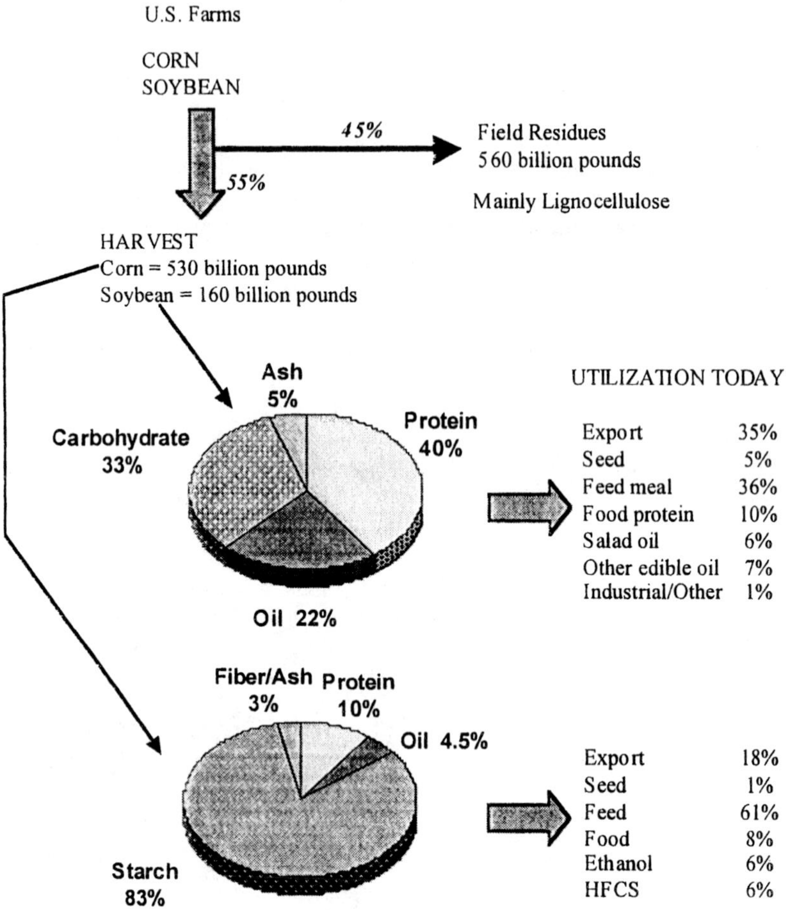

Figure 4. Production, composition and current utilization estimates for corn and soybeans as examples of major U.S. crops.

other cases are still at the proof of concept stage. The following examples are indicative of the range of types of developments rather than being an all inclusive list.

Biomass remains a large factor due to the existing volume of annual production that arises from agricultural and forestry residues, and recycling of wastes. This lignocellulosic material contains much stored energy and molecular carbon structures. Approaches to utilize biomass are many-fold and include pyrolysis, enzymatic degradation/conversions, and thermochemical breakdown into molecular structures such as levulinate (*13*). There are even projects that attempt to increase overall production of lignocellulose through the introduction of new species, such as switchgrass (*Panicum virgatum* L.), giant reeds (*Miscanthus x giganteus*), and fast growing woody plants (e.g. based on Alder, Willow, or Poplar species) (*13*).

One of the most widely recognized renewable resources is ethanol which is used as a fuel octane enhancer, and can be blended with gasoline instead of other oxygenates such as methyl tert-butyl ether (MTBE). Ethanol can also be used as a basic molecular building block. The primary source of U.S. ethanol is corn (Figure 4), generated via a fermentation process carried out within corn processing biorefineries. In 1998, 30 billion pounds of corn grain was used to produce 1.3 billion gallons of ethanol – reducing the demand for imported oil by 100,000 barrels per day (*14*). Improvements in the enzymatic degradation of lignocellulose would make additional use of crop stover, and forestry residues, more economically attractive.

The majority of soybean protein is used in feed while over 95% of the oil is used as direct edible vegetable oil or in baking/margarine (Figure 4). The small amount of industrial use of soybean oil is a reflection of the functionality being more suited to edible use because of high levels of oleic (C18:1) fatty acids, and relatively low levels of the more reactive fatty acids. However, industrially useful fatty acids do exist in several plant species. For example, some cultivars of oilseed rape and crambe have high levels of erucic acid (C22:1), castor seeds are a source of ricinoleic acid (C18:1 with a hydroxy group), and coriander is high in petroselenic acid (a special C18:1 fatty acid that can be cleaved to provide adipic and lauric acids). By using genomics and transgenic methods it is possible to isolate the key genes and move these to high oil producing plants like soybeans and canola. DuPont has achieved the transfer, to oil crops, of a gene from *Vernonia* that allows the production of an epoxide functional group on the fatty acid chain, opening up many new industrial uses. Renewable sources for lubricants, coatings, surfactants, plastics, and reactive oils are clearly on the horizon.

Polylactic acid (PLA) has been known for many years but use of this biodegradable "plastic" polymer was largely confined to internal surgical stitching due to the relatively high cost. Recent advances in the process for fermentation of corn starch to produce lactate, coupled with chemical lactide formation, have resulted in a cost that is economically attractive. Cargill Dow Polymers have announced investments of over $300 million in a new facility in

Nebraska that will produce 140,000 metric tons PLA per year. Corn will be the major renewable source for raw inputs, and the PLA resins will be manufactured into plastics for packaging and fibers.

Polyhydroxybutyrate (PHB) is a polyhdroxyalkanoate polymer with plastic-like properties that is synthesized by some soil bacteria (e.g. *Alcaligenes eutrophus*). The genes for the pathway have been cloned and inserted into plants with successful production of PHB. Additional research is required to optimize the properties: the butyrate-valerate copolymer has superior functionality but it remains more difficult to express this material in plants compared to the pure polymer (*15*). Improved expression levels and localization would also assist in separations and in the economic cost to produce this renewable biodegradable plastic.

It has been discovered that spider silk proteins have extraordinary tensile strength relative to their weight (*16*), and could form the basis of a whole new set of structural materials and novel fibers. The genes have been cloned and the proteins expressed in the milk from transgenic goats, providing sufficient supply to begin functional evaluations. The spider proteins have particular 3-dimensional folding and are rich in alanine and glycine amino acids, suggesting that design of novel oligopeptide structures with particular functions will be possible in the foreseeable future. Other areas of nature are rich in materials that have not yet been utilized for everyday products. Sea shells contain only 1-3% protein but it appears that their strength derives mainly from the properties of that protein (*17*). Other marine organisms could also be a useful source of new genes that code for interesting substances.

Conclusions

From traditional lignocellulosic biomass to genes from extreme-environment organisms, the research around renewable resources is beginning to show exciting promise for future years. However, this is not the time to relax but rather it is an appropriate time to push forward with aggressive new research into projects that have challenging goals. Incremental linear progress is inadequate when we need breakthrough type achievements: only "big science" will suffice to fix the big problems of future sustainable production.

Research in specific areas alone will not be sufficient to meet the needs within the timeframe available. History demonstrates that research results can take 15-20 years to translate into commercial realities – mainly, due to the serial approach typical of the traditional R&D system. Alternatively, parallel approaches can reduce the time from discovery to commercial products by many years. Integration across the various areas required to turn science into technology can lead to better products, more successful products, and a remarkable reduction in time to market.

The roadmap for renewable resources (*2*) identifies the key areas where integration would assist in achieving a significant increase in the use of renewables within the next 20 years. The contention is that brilliant research in one particular area is not enough *per se*. For example, assume that genomics provides a gene for a novel material that can replace the 60 billion pounds of plastics used in the U.S., and that this can be produced in corn stover with no incremental cost to existing production. This sounds wonderful. However, it is not of great benefit if the material is intricately tied to cellulose and cannot be extracted, nor separated, from the stover without the need for excessive amounts of energy. Future consumer value will best arise from a combination of simultaneous progress in areas such as plant science, production, processing, and materials utilization (Figure 5).

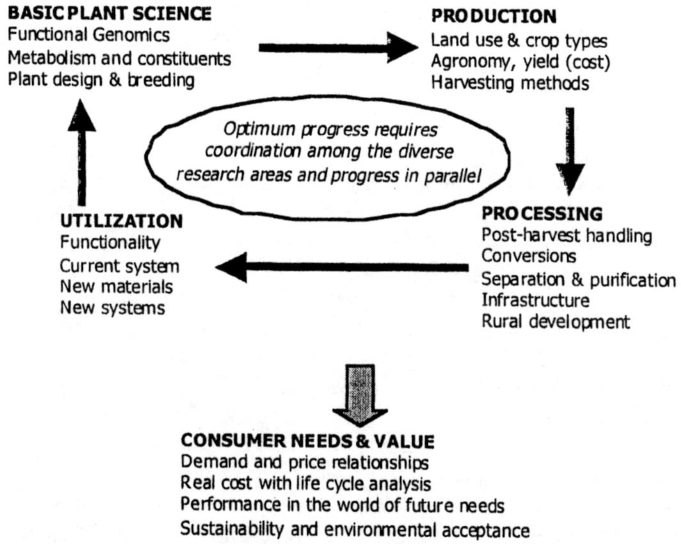

Figure 5. Areas of integration for maximum progress toward renewable resources with real recognizable value.

If the optimum value arises from a combination of these aspects, then it follows that research should focus on multi-aspect projects. Thus, most support should be directed toward those projects that include integrated concepts. While fossil fuels will still provide the majority of inputs for material goods over the next 20 years, we cannot be delinquent in developing renewable resources. The technology required to achieve sustainable development will not just emerge, and it will certainly not arise from emotional demonstrations concerning environmentalism. As has been demonstrated in many previous eras, solid, well-managed science has the ability to deliver. The required support for scientific progress around renewable resources is just beginning to emerge both in the private and public sectors. The future is indeed exciting and we can dream of sustainable well-being for our children.

References

1. *Plant/Crop-Based Renewable Resources 2020.* DOE/GO-10098-385. U.S. Department of Energy: Washington, D.C., 1998.
2. *The Technology Roadmap for Plant/Crop-Based Renewable Resources 2020.* DOE/GO-10099-706. U.S. Department of Energy: Washington, D.C., 1999.
3. Lugar, R.G.; Woolsey, R.J. *Foreign Affairs* **1999**, 78, 88-102.
4. *Climate Change – The IPCC Scientific Assessment.* Intergovernmental Panel on Climate Change: Cambridge University Press, UK, 1990.
5. Layman, P.L. *Chem. Eng. News* p. 20, October 1999.
6. Reisch, M.S. *Chem. Eng. News,* p. 33, February 2000.
7. *Annual Energy Review.* Energy Information Administration, U.S. Department of Energy: Washington, D.C., 1999.
8. Bozell, J.J.; Landucci, R. *Alternative Feedstocks Program: Technical and Economic Assessment, Thermal/Chemical and Bioprocessing Components.* U.S. Department of Energy: Washington, D.C., 1993.
9. Rooney, T. *Lignocellulosic Feedstock Resource Assessment.* Report SR-580-24189, National Renewable Energy Laboratory, Golden, CO, 1998
10. *Biobased Industrial Products.* National Research Council, National Academy Press, Washington, D.C., 2000.
11. McLaren, J.S. *Pesticide Outlook* December **1998**, 36-41.
12. *National Plant Genome Initiative Progress Report.* National Science and Technology Council, Interagency Working Group on Plant Genomes, Office of Science and Technology Policy, Washington, D.C., 1999.
13. *Biomass: A Growth Opportunity in Green Energy and Value-Added Products.* Overend, R.P.; Chornet, E., Eds. Proceedings of the Fourth Biomass Conference of the Americas: Elsevier Science, 1999.
14. *The World of Corn 1999.* National Corn Growers Association, St. Louis, MO.
15. Slater, S.; Mitsky, T.A.; Houmiel, K.L.; Hao, M.; Reiser S.E.; Taylor, N.B.; Tran, M.; Valentin, H.E.; Rodriguez, D.J.; Stone, D.A.; Padgette, S.R.; Kishore, G.; Gruys, K.J. *Nature Biotechnology* **1999**, *17*, 1011-1016.
16. Tirrell, D. *Science* **1996**, *271*, 39-40.
17. Smith, B.L. *Chem. Ind.* **1998**, *16*, 649-653.

Chapter 4

Synthesis of δ-Aminolevulinic Acid

Luc Moens

National Renewable Energy Laboratory (NREL), 1617 Cole Boulevard, Golden, CO 80401

Improvements in the industrial production of levulinic acid from lignocellulosic materials have revived the interest in the use of this renewable chemical as a 'platform chemical' for the chemical industry. Here, an improved synthetic method is presented for the production of δ-aminolevulinic acid (DALA) from levulinic acid. DALA is attracting much attention as an environmentally benign pesticide, and antitumor agent.

The oil refinery as we know it today has made it possible for the chemical industry to get immediate access to a wide variety of hydrocarbons that serve as starting material or 'platform chemicals' for a plethora of chemicals and materials which have become the basis for the production of most consumable goods in our society. However, petroleum is a non-renewable resource and, as we witnessed during the 1970's, its supply can easily become jeopardized by global socio-political events. In addition, the use of fossil resources to produce energy and chemicals is one of the primary causes of carbon dioxide buildup in the atmosphere that is believed to cause global climate changes. This has caused a renewed global interest in the use of lignocellulosic biomass as a renewable feedstock for energy and chemicals. However, the chemical structures that make up biomass are much more complex, and this creates a challenge for the chemical industry. Thus, entirely new chemical processes must be developed that convert the complex lignocellulosics into chemicals. In analogy with the oil refineries that specialize in the 'cracking' of petroleum into smaller molecules, one can speak of a 'biorefinery' wherein biomass is broken down into well defined molecules that can be used for further processing to other chemicals. This concept has already been realized to some extent in the case of sugar and alcohol production, as well as specialty chemicals such as furans. However, by no means is the concept as advanced as the 'state of the art' in a current petroleum refinery. Key to the success of a biorefinery will be the progress in three different areas: a) a better understanding of the physico-chemical behavior of the many available biomass feedstocks in process reactors, b) the development of new chemical technologies that can deal with the complex and heterogeneous nature of biomass, and c) a better understanding of the reactivities and secondary condensation reactions of the reaction intermediates and products that are formed during the chemical conversion process. Especially the latter is an important requirement for biomass processing, since the intermediates and products often are very reactive due to their higher degree of oxygenation compared to petrochemicals.

One of the unique chemicals that can be produced from the cellulosic fraction of biomass is levulinic acid. For a long time, this compound has been known as a by-product of the acid hydrolysis of cellulose or monomeric hexoses such as glucose, but until recently, no good method existed for producing levulinic acid in high yields through this type of reaction (*1,2*). The acid-catalyzed degradation of e. g., cellulose first generates glucose, which undergoes further dehydration to form 5-hydroxymethylfurfural (5-HMF). The latter is known to be very unstable under the acidic conditions, and tends to hydrolyze further with formation of levulinic acid (C_5-unit) and formic acid (C_1-unit) (Scheme 1).

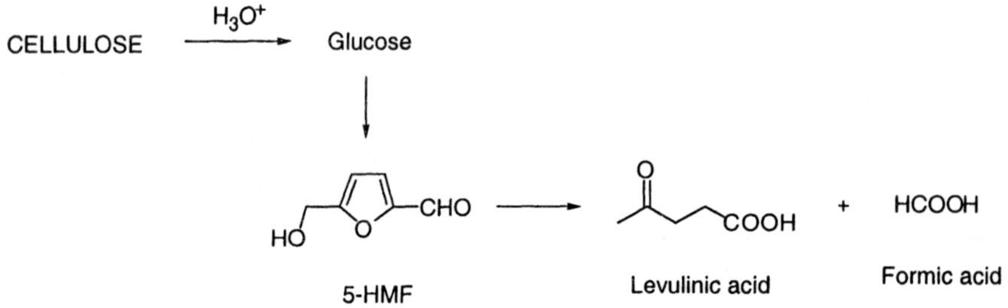

Scheme 1. Formation of levulinic acid from cellulose through acid hydrolysis

Along this pathway, a number of condensation reactions can take place, and it has been found that when the levulinic acid is not removed quickly from the reaction medium, tarry materials are formed that drastically lower the yield of the levulinic acid. Historically, this has kept the cost of levulinic acid too high to make this compound attractive as a 'platform chemical' for the production of a series of derived chemicals and materials. However, recent work at Biofine, Inc. (Massachusetts), led to the development of an optimized process that generates levulinic acid in yields of 70-90% starting from waste paper (*3*). Economic evaluations of the new process indicated a much lower cost for levulinic acid, and this bodes very well for the expanded use of this chemical in the very near future. Today, levulinic acid is primarily produced starting from maleic acid, and serves as a catalyst in the manufacture of specialty adhesives, rubber, pharmaceutical, plastic, and synthetic fiber products. It is predicted that the cheaper biomass-derived levulinic acid will make it more suitable as a platform chemical for the production of many new chemicals and materials.

This chapter deals with the synthesis of δ-aminolevulinic acid (DALA), and an improved synthetic method starting from levulinic acid will be presented that was recently developed in our laboratory.

Use of δ-Aminolevulinic Acid (DALA) as a Herbicide

DALA is a naturally occurring substance present in all plant and animal cells. It serves as the precursor molecule in the cellular production of tetrapyrroles such as chlorophyll and the heme of hemoglobin, and thus it plays a key role in such vital processes as photosynthesis and oxygen transport (*4,5*). Interestingly, DALA can turn into a powerful herbicide when it is applied externally, in which case the metabolic balance is disturbed and an excess of tetrapyrrole intermediates builds up within the plant cells. Under light exposure, i. e., as soon as the plants are exposed to daylight, the accumulated tetrapyrroles use the sunlight to convert oxygen into singlet oxygen that kills plant cells by excessively oxidizing the cell material (*6*). DALA has also been found to have insecticidal (*7,8*) and antitumor properties (*9*). Because DALA is completely metabolized, no unnatural residues remain. DALA that is not taken up by the plants decomposes within a day or two, so toxicity and impact on the environment are minimal. Under greenhouse conditions, DALA is only effective on dicotyledons or broadleaf plants. Monocotyledons such as grasses and grains are able to tolerate it. As such, DALA may be ideal for killing broadleaf weeds in grass such as in turf farms, golf courses, and residential lawns. Similarly, it can be used for monocotyledon crops such as corn, wheat, oats, and barley. For both of these uses, DALA requires the addition of small amounts of chemical "modulators" to trigger its herbicidal action. By varying the modulator, DALA can be formulated such that a variety of weeds are killed while leaving grass or crops unharmed. Unfortunately, the production cost for DALA has been very high, and this has prevented its widespread application. With the advent of cheaper levulinic acid through the Biofine process, the development of better synthetic routes towards DALA may lead to its commercialization.

Overview of Known Synthetic Pathways for DALA Production

Since the 1950's, the synthesis of DALA has undergone continuous improvements and a variety of starting materials have been tested, including levulinic acid. The following is an overview of successful synthetic approaches that have been described in both the scientific and patent literature. Note that throughout the text the acronym DALA will be used for both the hydrochloride salt and the free DALA. The free DALA is a very unstable compound (1,2-aminoketone) that undergoes spontaneous dimerization to form a stable pyrazine **2** after oxidation of the intermediate dihydropyrazine **1** (Figure 1) (*10*).

The initial reports described routes that involved a sequential build-up of the carbon chain, combined with the introduction of the amino group through the use of potassium phthalimide (Gabriel synthesis) (*11-15*). The use of furans as starting material offered a route that was relatively simple, as was shown in the conversion of N-benzoylfurfurylamine (**3**) to DALA (Figure 2) (*16*).

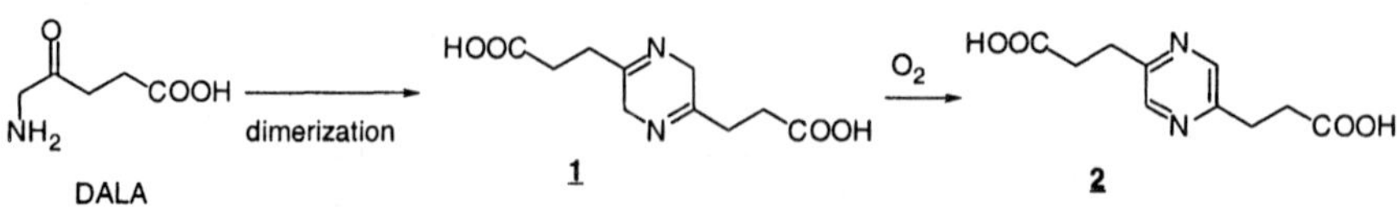

Figure 1. Dimerization of DALA to form a stable pyrazine. (Adapted with permission from reference 10. Copyright 1981 Japan Inst. Heterocyclic Chem.)

The application of the electrochemical variant of the Clauson-Kaas oxidation allowed the formation of the dihydrodimethoxyfuran **4**. Catalytic hydrogenation of this hemiacetal led to a tetrahydrofuran that after oxidation and hydrolysis resulted in the formation of DALA. A similar pathway, shown in Figure 3, is the oxidation of 5-acetamido-methylfurfural (**5**) with singlet oxygen (*17,18*). This generates a butenolide (**6**) that after selective reduction with borohydride, is converted into a furan-2(5H)-one **7**. Zinc powder, activated under sonication in acetic acid, can cause ring opening of the butenolide ring.

An analogous sequence was patented that applied this photochemical protocol to the phthalimide form of furfural. In a more recent report, ruthenium trichloride was used as an oxidizing agent in the ring opening of the phthalimide of tetrahydrofurfurylamine(*19,20*). Closely related to this work is the oxidative ring opening of amine **8** to DALA with chromium oxide in sulfuric acid (Figure 4) (*21*).

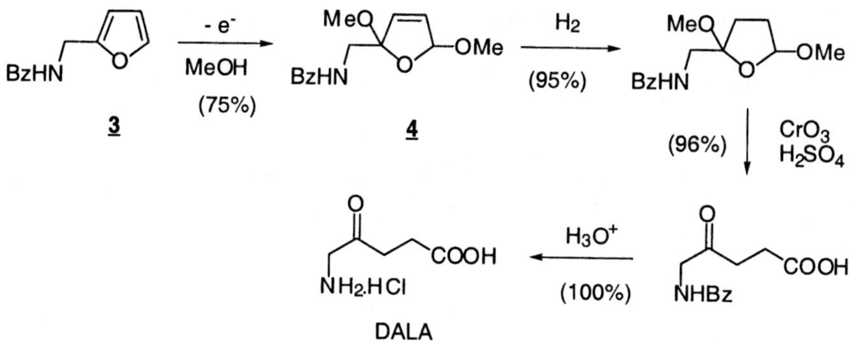

Figure 2. Conversion of furfurylamine into DALA via anodic oxidation. (Adapted with permission from reference 16. Copyright 1958 RSC.)

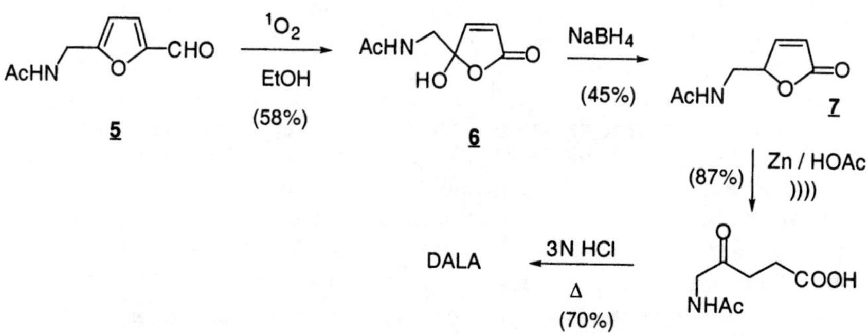

Figure 3. Synthesis of DALA from furfurylamine using singlet oxygen. (Adapted with permission from reference 17. Copyright 1995 Thieme)

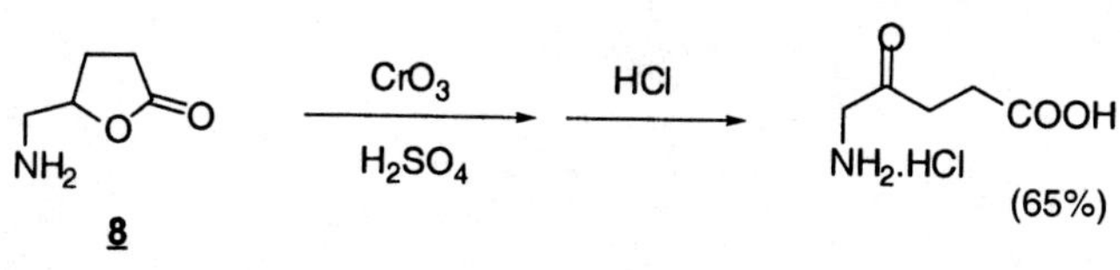

Figure 4. DALA synthesis via oxidation of a tetrahydrofuranone. (Adapted with permission from reference 21. Copyright 1998 RSC.)

Another precursor for DALA is succinoyl chloride monomethyl ester that can undergo chain extension in the presence of hippuric acid and a base catalyst such as γ-picoline (Figure 5) (*22*). The initial reaction involves acylation of the dianion of hippuric acid. The resulting oxazolinone **9** can be hydrolyzed in aqueous hydrochloric acid to yield DALA. In a similar manner, succinic anhydride can serve as acylating agent when it is treated with the dianion of ethyl hippurate (*23*).

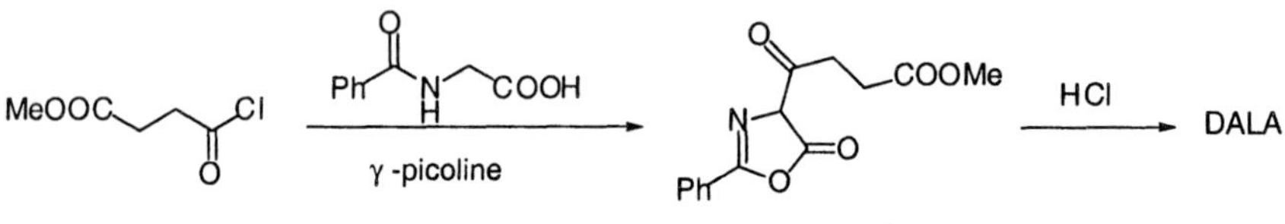

9

Figure 5. DALA synthesis via acylation of hippuric acid. (Reproduced from reference 22.)

The need for ^{13}C-labeled analogues of DALA for the elucidation of biosynthetic and metabolic pathways towards heme and chlorophyll in living cells led to the development of other synthetic routes. The reaction of succinoyl chloride monoesters with ^{13}C-labeled potassium cyanide provided a straightforward route towards acyl cyanides that are convenient precursors for DALA (*24-27*). More recently, a versatile and efficient syntheses of ^{13}C-4-, ^{13}C-5- and ^{15}N-DALA was reported (*28*). The pathway for these syntheses was based on a chain elongation of isotopically labeled glycines.

Another strategy, shown in Figure 6, involved the hydrolytic ring opening of lactam **11** that was obtained by hydrogenation of 5-hydroxy-2-pyridone (**10**) (*29,30*). Lactams analogous to **11** were found to be accessible from N-methoxycarbonyl piperidine and provided an interesting but longer synthetic route to DALA (*31*).

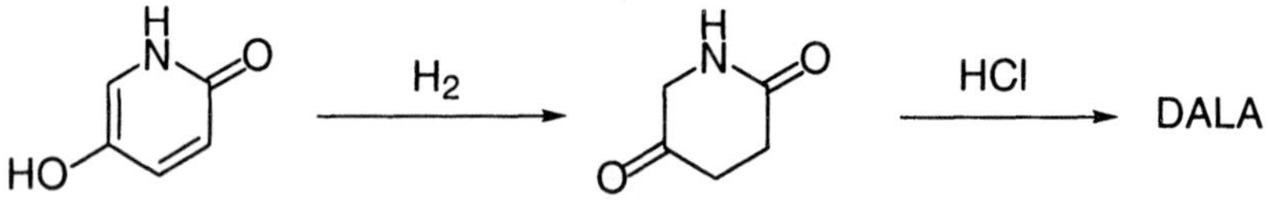

10 11

Figure 6. DALA synthesis via ring opening of a lactam. (Adapted with permission from reference 29. Copyright 1986 Thieme.)

Over the last 50 years, only a few papers have appeared in the literature that describe the use of levulinic acid as starting material for the production of DALA. The selective introduction of an amino group in the C5-position represents a major challenge because all the hydrogen atoms in levulinic acid are activated by a neighboring carbonyl group. This can cause enolization at different sites on the carbon chain, resulting in low selectivity and/or side reactions. One of the first successful attempts to produce DALA from methyl levulinate involved the removal of the methylene hydrogens *via* a bromination and dehydrobromination sequence to form methyl acetylacrylate (12) (Figure 7) (*32*). The activated olefinic bond was then exploited to 'direct' the subsequent amination step towards the C5-position. Treatment of 12 with ethyl nitrite

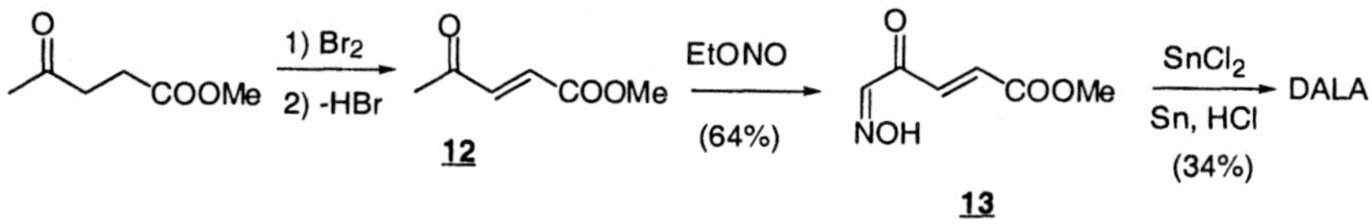

Figure 7. DALA synthesis through selective activation of C5 in levulinic acid

generated oxime 13, in which both the C=N and C=C bond could be selectively reduced using tin and stannous chloride in an acidic medium. A later report indicated that 4,5-dioxovaleric acid can undergo a non-enzymatic transamination reaction with glycine (*33*). However, by far the most straightforward activation of the C5-position involves direct bromination of levulinic acid or its esters (*34,35*). This halogenation is not very selective and will be discussed in more detail later in this chapter as part of the discussion of our own work. Generally, the bromination generates mixtures of 3-bromo- and 5-bromolevulinate which can be separated through careful distillation. The 5-bromolevulinate 14 can then undergo nucleophilic substitution reactions with nitrogen-based reagents. As shown in Figure 8, a very efficient reagent is

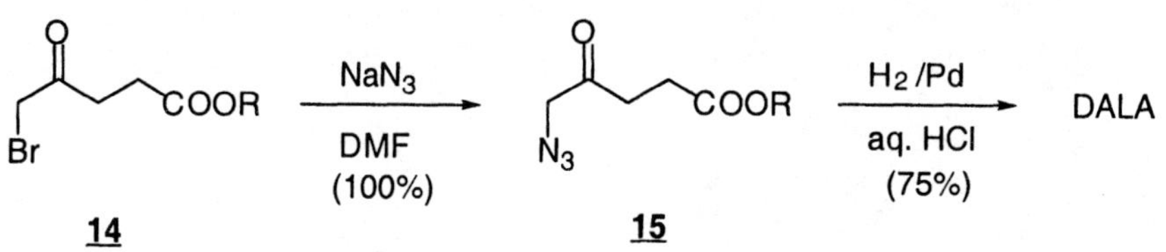

Figure 8. DALA synthesis via nucleophilic amination. (Adapted with permission from reference 34. Copyright 1994 Marcel Dekker, Inc.)

sodium azide that affords azide **15** in quantitative yield. The latter was hydrogenated to DALA over palladium catalysts in aqueous hydrochloric acid. In a more recent report, the sodium azide was replaced by potassium phthalimide to effect a Gabriel-type amination reaction (*36*). This well established protocol generated DALA after hydrolysis of the intermediate phthalimide in aqueous hydrochloric acid.

Improved Synthetic Pathway to DALA

Due to the availability of cheap levulinic acid through the process developed at Biofine, Inc., we decided to improve the existing pathways for converting this feedstock into DALA. The challenge that we faced at the onset of the project was twofold: first it was necessary to find a reagent that could activate the C5-position selectively without affecting the other sites on the levulinic acid molecule. Secondly, we had to find a reagent for the amination step that would be more 'atom economical', i.e., one that would not generate waste materials with large amounts of carbon in its structure. For instance, the use of potassium phthalimide results in the generation of phthalic acid as a by-product, which contains eight carbon atoms. We wanted to avoid the formation of such 'expensive' waste by searching for smaller but highly reactive aminating agents. On the other hand, we did not want to have to deal with small molecules such as e.g., sodium azide that is highly toxic.

At this point, we had the choice between two strategies that are illustrated in Scheme 2. The first pathway would involve deprotonation of levulinic acid (or ester) at the C5-position to generate an anion that can react with an electrophilic aminating agent to form the C-N bond. The second approach would be to carry out a specific halogenation at C5, followed by a nucleophilic amination.

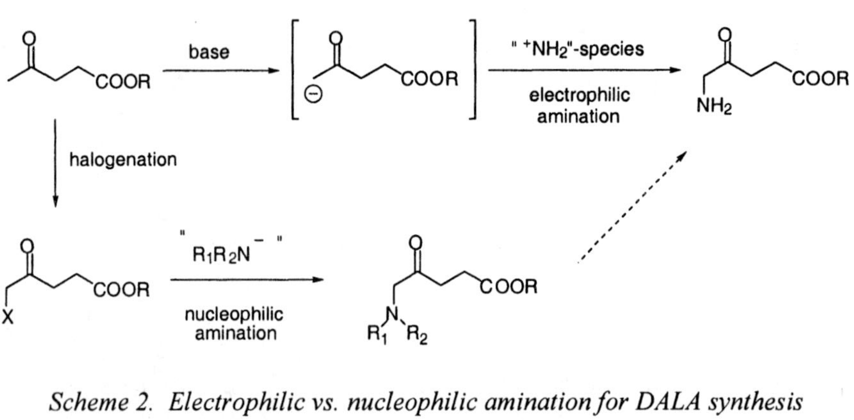

Scheme 2. Electrophilic vs. nucleophilic amination for DALA synthesis

Activation of the C5-Position

Because the selective halogenation step appeared to be less attractive, we decided to explore the electrophilic amination pathway first. A review of the literature revealed that the deprotonation of levulinic acid with 2 equivalents of strong base such as e. g. lithium diisopropylamide (LDA), generates a dianion that is insoluble in most solvents (*37*). Consequently, the reactivity of the dianion is very low and leads to poor yields when for instance acylating agents are added. Another issue that we had to keep in mind with respect to deprotonation of levulinic acid was that it has a tendency to undergo spontaneous cyclization and dehydration at elevated temperatures, with formation of α-angelica lactone (Figure 9) (*2*). This suggested that the best strategy would be to use a strong and bulky base that can effect kinetic deprotonation at the C5 position (as opposed to C2 or C3). The resulting kinetic enolate could then, at least in theory, react with an electrophilic aminating agent.

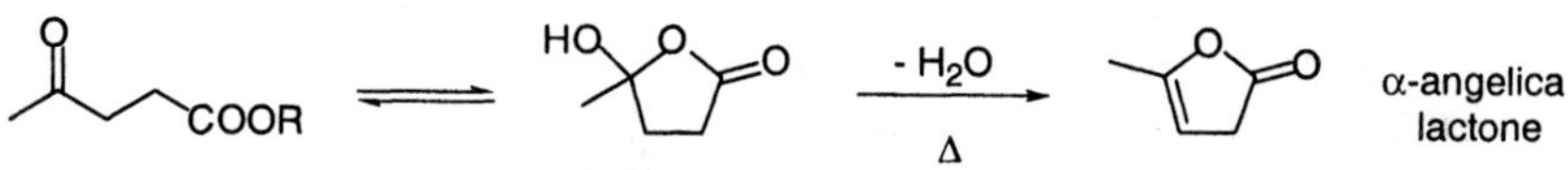

Figure 9. Spontaneous cyclization and dehydration of levulinic acid.

The chemical literature contains several examples of very efficient electrophilic aminating agents, and a representative number of such reagents is shown in Figure 10 (*37-43*).

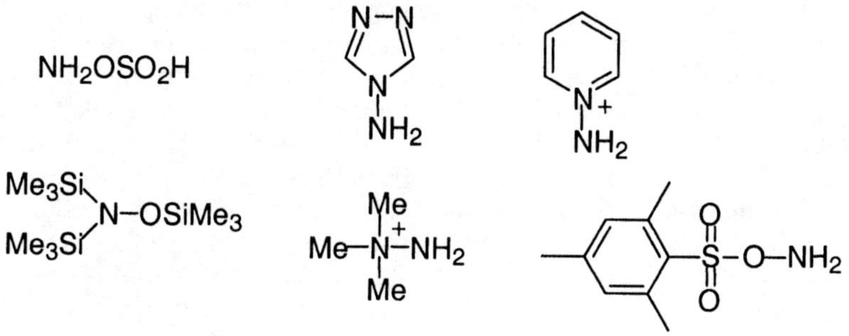

Figure 10. Examples of known electrophilic aminating agents

However, none of these reagents seemed to be appropriate for our DALA production because they are either very toxic, water-sensitive or unstable, and have the added disadvantage of low 'atom efficiency' as they generate waste that contains molecular residues of significant size. Because of these factors, we did not spend much time exploring the reactivity of levulinic acid under these conditions. However, a future detailed study of enolate chemistry with levulinic acid using a variety of bulky bases would be worthwhile because of the high functionality of this ketoacid.

Bromination of Levulinic Acid

Considering that we needed to find 'simple' reaction conditions and inexpensive reagents for a commercially viable DALA synthesis, we decided to revisit the nucleophilic amination route (Scheme 2). It was particularly intriguing to us that the bromination of levulinic acid had been used by several authors, but that relatively little had been done to optimize this reaction to where better bromination of C5 vs. C2 or C3 could be achieved (*44-49*). This led us to believe that improvements could be made to the preparation of 5-bromolevulinic acid (or ester) by careful control of the reaction conditions. The literature suggests that the 5-bromo/3-bromo isomer ratio is highly dependent on the solvent. For instance, in a polar medium such as methanol, the ratio is 3 : 1 (favoring 5-bromo isomer **14**), while in ether only the 3-bromo isomer is formed. Since the bromination is induced by traces of acid (HBr) that catalyze enolization of the keto group *(50)*, the 5-bromo ester is probably formed first (kinetic product), whereafter it slowly isomerizes to the 3-bromo isomer (**16**) under the influence of the dissolved HBr. An important by-product from the bromination reaction is the 3,5-dibromo compound **17** as well as residual levulinic acid (or ester). When the bromination is carried out in methanol, one isolates methyl bromolevulinates as a result of HBr-catalyzed esterification (*44*). After considerable experimentation with different concentrations of methyl levulinate and different rates of bromine addition, we started to notice that the 5-bromo/3-bromo ratio could vary considerably. We found that the ratio was usually close to 3 : 1 (as measured by ^{1}H-NMR) immediately after termination of the bromination reaction, but that it sometimes decreased to only 1 : 1 after the work-up. The work-up consisted of removing all traces of solvent, HBr and residual bromine under reduced pressure at room temperature (using a rotary evaporator). Since this decreased ratio represented a serious loss of the desired 5-bromo ester, we tried to modify the protocol such that the initial 3 : 1 ratio would be preserved throughout the work-up procedure. We discovered two protocols that seemed to maintain the 3 : 1 ratio: 1) carrying out the bromination under reflux, and b) carrying out the bromination at room temperature, but removing the methanol and traces of HBr by slow distillation at atmospheric pressure. We believe that these two modes help the rapid evaporation of HBr before it starts to isomerize the product mixture. If this is true, then the isomerization is dependent on the concentration of the HBr. When the reaction solution is being concentrated at room temperature using a rotary evaporator, some of the HBr may remain in solution long enough to start the isomerization process at the gradually increasing product concentrations as the solvent evaporates. These findings corroborate earlier studies (*45-47*) in that the selectivity of the bromination reaction is controlled by an equilibrium reaction that is supported by the presence of HBr in the solution. We succeeded in obtaining fairly consistent product ratios that remained close to the one indicated in Scheme 3.

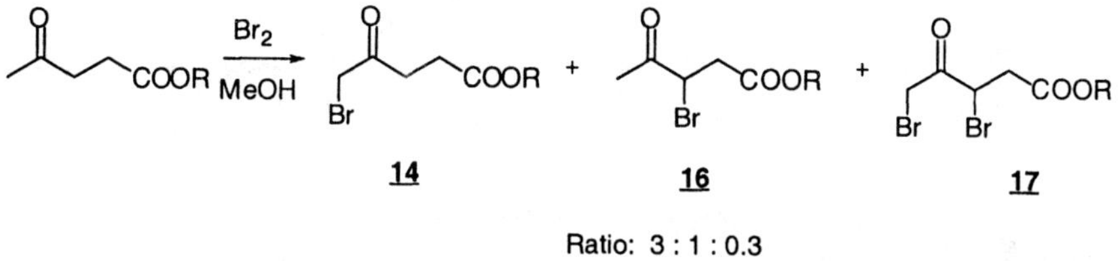

Scheme 3. Best ratio obtained for the bromination reaction

This product ratio is also in agreement with those observed in the literature. However, we have not investigated what the real steady state concentration of each product is when full equilibrium is reached.

A more disturbing observation, however, that was made during the early studies of the bromination of levulinic acid, was that during the vacuum distillation of the crude product mixture the boiling points of the isolated fractions fluctuated erratically (*44*). This made it difficult to get pure samples of each isomer, and each distillation fraction had to be monitored carefully. We made similar observations during our work on this bromination reaction, and we started to suspect that traces of HBr caused isomerization of the brominated levulinate mixture during the distillation process. Such traces could possibly be generated under the high-temperature conditions inside the distillation pot. Thus, we reasoned that addition of small amounts of weak base might prevent the build-up of the acid catalyst. When the crude brominated product mixture was treated with weak bases such as calcium hydroxide or sodium bicarbonate, the reaction mixture underwent chemical changes and only unidentified products were observed by NMR. However, our hypothesis regarding the influence of traces of HBr was confirmed when calcium carbonate was added: the crude brominated products remained unaffected and the subsequent distillation of the turbid suspension proceeded much more smoothly without the erratic boiling point fluctuations. We tested this new approach on a 50 – 100 g scale, and while we have not yet optimized this procedure, it did provide us with sufficient methyl 5-bromolevulinate for further work on the subsequent amination reaction. Future work will hopefully support the notion that this protocol is equally effective at kilogram scales.

Development of an Improved Amination Step

As mentioned earlier, the aminating agent would ideally contain the least amount of carbon atoms possible in order to minimize the quantity of waste after completion of the reaction. After a detailed literature study, we came across reports that described the use of sodium diformylamide (**18**) as a nucleophilic aminating agent (*51-53*). This reagent is readily prepared by mixing sodium methoxide and formamide at room temperature, and the resulting sodium diformylamide is formed as a precipitate that can be isolated as a white solid. This reagent has excellent reactivity as a nucleophilic nitrogen species towards halogenated compounds (*51-57*), and we were successful in using this reagent for our amination step with methyl 5-bromolevulinate (Figure 11). After hydrolysis of the intermediate diformylamide **19**, we obtained DALA in quantitative yields with >90% purity (*58*). The purification of the DALA will have to be finetuned in future scale-up work, but we found that this crude DALA proved to be active as a herbicide during initial testing in a greenhouse environment. Another positive aspect of this process is that sodium diformylamide can be used in a variety of solvents, such as e. g. acetonitrile,

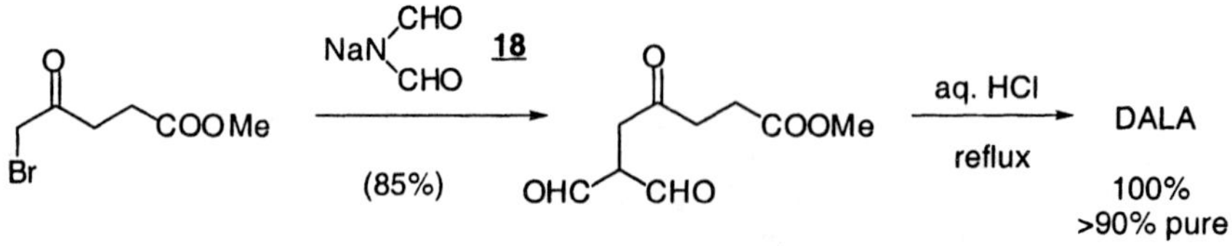

Figure 11. New DALA synthesis from levulinic acid with sodium diformylamide as nucleophilic aminating agent. (Reproduced from reference 58.)

tetrahydrofuran, methyltetrahydrofuran and methyl formate. Although this reagent certainly has the advantage of being a small molecule, it still generates sodium salts as part of the waste stream. We are therefore continuing our search for small nucleophilic nitrogen species that avoid the formation of such inorganic waste. Although we have identified such a reagent, this work is still proprietary at the time of this writing, and further details will be disclosed in due course.

Acknowledgment

This work was supported by the Office of Industrial Technologies (U. S. Department of Energy), Biofine, Inc.; Chemical Industry service, Inc.; and the New York Energy Research and Development Agency. I am also grateful to Dr. Joseph J. Bozell at NREL for the many stimulating discussions.

References

1. Leonard, R. H. *Ind. Eng. Chem.* **1956**, *48*, 1331.
2. Kitano, M.; Tanimoto, F.; Okabayashi, M. *Chem. Econ. Eng. Rev.* **1975**, *7*, 25.
3. Fitzpatrick, S, F. U.S. Patent 5,608105, 1997.
4. Shemin, D.; Russell, C. S.; Abramsky, T. *J. Biol. Chem.* **1955**, *215*, 613.
5. Leeper, F. J. *Nat. Prod. Rep.* **1989**, *6*, 171.
6. Rebeiz, C. A.; Montazer-Zouhoor, A.; Hopen, H. J.; Wu, S. M. *Enzyme Microb. Technol.* **1984**, *6*, 390.
7. *Porphyric Pesticides;* Duke, S. O.; Rebeiz, C. A., Eds.; American Chemical Society Symposium Series, Washington, DC, 1994, Vol. 559.
8. Rebeiz, C. A.; Juvik, J. A.; Rebeiz, C. C. *Pestic. Biochem. Physiol.* **1988**, *30*, 11.
9. Rebeiz, N.; Arkins, S.; Rebeiz, C. A.; Simon, J.; Zachary, J. F.; Kelley, K. W. *Cancer Res.* **1996**, *56*, 339.
10. Franck, B.; Stratmann, H. *Heterocycles* **1981**, *15*, 919.
11. Lartillot, S.; Baron, C. *Bull. Soc. Chim. France* **1966**, 3798.
12. Lartillot, S.; Baron, C. *Bull. Soc. Chim. France* **1964**, 783.
13. Tschudy, D. P.; Collins, A. *J. Org. Chem.* **1959**, *24*, 556.
14. Neuberger, A.; Scott, J. J. *J. Chem. Soc.* **1954**, 1820.
15. Shemin, D.; Russell, C. S. *J. Am. Chem. Soc.* **1953**, *75*, 4873.
16. Marei, A. A.; Raphael, R. A. *J. Chem. Soc.* **1958**, 2624.
17. Cottier, L.; Descotes, G.; Eymard, L.; Rapp, K. *Synthesis* **1995**, 303.
18. Takeya, H.; Shimizu, T.; Ueki, H. U.S. Patent 5,380,935, 1995.
19. Kawakami, H.; Ebata, T.; Matsushita, H. *Agric. Biol. Chem.* **1991**, *55*, 1687.
20. Ebata, T.; Kawakami, H.; Matsumoto, K.; Koseki, K.; Matsushita, H. U.S. Patent 5,284,973, 1994
21. Appleton, D.; Duguid, A. B.; Lee, S.-K.; Ha, Y.-J.; Ha, H.-J.; Leeper, F. J. *J. Chem. Soc., Perkin Trans. 1*, **1998**, 89.
22. Aronovo, N. I.; Makhova, N. N.; Zavyalov, S. I.; Volkenshtein, J. B.; Kunitskaya, G. M. U.S. Patent 3,846,490, 1974.
23. Evans, D. A.; Sidebottom, P. J. *J. Chem. Soc. Chem. Commun.* **1978**, 753.
24. Pfaltz, A.; Anwar, S. *Tetrahedron Lett.* **1984**, *25*, 2977.

25. Vishwakharma, R. A.; Balachandran, S.; Alanine, A. I. D.; Stamford, N. P. J.; Kiuchi, F.; Leeper, F. J.; Battersby, A. R. *J. Chem. Soc. Perkin Trans. 1*, **1993**, 2893.
26. Kurumaya, K.; Okazaki, T.; Seido, N.; Akasak, Y.; Kawajiri, Y.; Kajiwara, M. *J. Lab. Comp. Radiopharm.* **1988**, *27*, 217.
27. Battersby, A. R.; Hunt, E.; McDonald, E.; Moron, J. *J. Chem. Soc. Perkin Trans.1*, **1973**, 2917.
28. Wang, J.; Scott, A. I. *Tetrahedron Lett.* **1997**, *38*, 739.
29. Herdeis, C. *Synthesis* **1986**, 232.
30. Herdeis, C.; Dimmerling, A. *Arch. Pharm.. (Weinheim)* **1984**, *317*, 304.
31. Matsumura, Y.; Takeshima, Y.; Okita, H. *Bull. Chem. Soc. Japan* **1994**, *67*, 304.
32. Wynn, R. W.; Corwin, A. H. *J. Org. Chem.* **1950**, *15*, 203.
33. Beale, S. I.; Gold, M. H.; Granick, S. *Phytochemistry* **1979**, *18*, 441.
34. Ha, H.-J.; Lee, S.-K.; Ha, Y.-J.; Park, J.-W. *Synth. Commun.* **1994**, *24*, 2557.
35. Metcalf, B. W.; Adams, J. L. U.S. Patent 4,325,877, 1982.
36. Benedikt, E.; Köst, H.-P. *Z. Naturforsch., B: Anorg. Chem. Org. Chem.* **1986**, *41B(12)*, 1593.
37. Murray, W. V.; Wachter, M. P. *J. Org. Chem.* **1990**, *55*, 3424.
38. Erdik, E.; Ay, M. *Chem. Rev.* **1989**, *89*, 1947.
39. Wallace, R. C. *Aldrichim. Acta* **1980**, *13*, 3
40. Casarini, A.; Dembech, P.; Lazzari, D.; Marini, E.; Reginato, G.; Ricci, A.; Seconi, G. *J. Org. Chem.* **1993**, *58*, 5620.
41. Tamura, Y.; Minamikawa, J.; Ikeda, M. *Synthesis* **1977**, 1.
42. Andreae, S.; Schmitz, E. *Synthesis* **1991**, 327.
43. Gösl, R.; Meuwsen, A. *Chem. Ber.* **1959**, *92*, 2521.
44. MacDonald, S. F. *Can. J. Chem.* **1974**, *52*, 3257.
45. Rappe, C. *Arkiv. Kemi* **1962**, *20*, 51.
46. Rappe, C. *Arkiv. Kemi* **1959**, *14*, 467.
47. Rappe, C. *Arkiv. Kemi* **1959**, *13*, 425.
48. Hurd, C. D.; Ferraro, J. R. *J. Org. Chem.* **1951**, *16*, 1639.
49. Overend, W. G.; Turton, L. M.; Wiggins, L. F. *J. Chem. Soc.* **1950**, 3500.
50. De Kimpe, N.; Verhé, R. In *The Chemistry of a-haloketones, a-haloaldehydes and a-haloimines;* Patai, S.; Rappoport, Z., Eds.; J. Wiley & Sons, New York, NY, 1988.
51. Yinglin, H.; Hongwen, H. *Synthesis* **1990**, 122.
52. Yinglin, H.; Hongwen, H. *Synthesis* **1990**, 615.
53. Ying-lin, H.; Hong-wen, H. *Tetrahedron Lett.* **1989**,5285.
54. Evans, J. C.; Goralski, C. T.; Rand, C. L.; Vosejpka, P. C. U.S. Patent 5,599,986, 1997.
55. Gramain, J. C.; Rémuson, R. *Synthesis* **1982**, 264.
56. Meduna, V.; Sawlewicz, P.; Vogt, R. *Synth. Commun.* **1989**, *19*, 1487.
57. Allenstein, E.; Beyl, V.; Eitel, W. *Chem. Ber.* **1969**, *102*, 4089.
58. Moens, L. U.S. Patent 5,907.058, 1999.

Chapter 5

Levulinate Esters from Biomass Wastes

Edwin S. Olson, Michelle R. Kjelden, Adam J. Schlag, and Ramesh K. Sharma

The Energy and Environmental Research Center, University of North Dakota, 15 North 23rd Street, Grand Forks, ND 58202-9018

The conversion of resin-bonded furniture and building waste to levulinate esters was investigated with a view to producing fuels, solvents, and chemical intermediates as well as other useful byproducts in an inexpensive process. The acid-catalyzed reaction of cellulosic materials with ethanol or methanol at 200°C gives good yields of levulinate and formate esters. A solid residue (charcoal) and a resinous lignin residue are also obtained. The excess alcohol solvent is recycled. An advantage of the reaction in alcohol is that wastewater is minimized and products are purified easily by distillation of the esters. Methanol and ethanol, which give the highest yields of levulinate esters, can be obtained at a low production cost. These esters are therefore the preferred intermediates for conversion to many other products. The ethyl ester has been converted in high yield to a number of other esters by Fischer esterification. The resulting ketoesters are substrates for a variety of condensation and addition reactions at the ester and keto groups. A market for these higher-value products is needed to justify the process cost.

Cellulosic Waste Processing

Cellulosic biomass constitutes a huge and renewable resource that can be converted to chemical and fuel feedstocks. More efficient means for conversions of agricultural, forest, and construction waste are sought so that useful biomass-derived products can compete with and eventually replace petroleum-based products, and at the same time reduce the disposal problems and pollution resulting from accumulation of cellulosic waste.

Some waste biomass is converted to useful products like ethanol in fermentation processes, and both fuel and chemical feedstocks have been prepared by acid-catalyzed thermal decomposition of cellulose (*1*). Heating lignocellulosic materials in aqueous acid hydrolyzes the cellulose to glucose and subsequently converts the glucose to levulinic (4-oxopentanoic) acid and formic acid. Advances in the thermal processing equipment have resulted in improvements in the conversion of cellulosics to levulinic acid (*2*). The potential usefulness of levulinic acid from this degradative processing is demonstrated by the hydrogenation of levulinic acid to the alternative fuel additive, methyltetrahydrofuran (methyl THF), which is currently being promoted for use in P-fuel (*3*). Although levulinic acid exhibits poor solubility, it does increase octane numbers (*4*). Soluble derivatives of levulinic acid could be developed for oxygenate fuel additives that would not only increase engine efficiency, but also lower emissions from the engines consuming these fuels. α-Angelica lactone is readily prepared by acid-catalyzed dehydration of levulinic acid and has been shown to increase octane ratings (*5*).

Several higher-value applications of the thermal degradation products of cellulosics are commercially developed or have been previously investigated. Furfural is derived from hemicellulose containing five-carbon sugars, and a large market exists for it as a chemical intermediate. Several years ago, diphenolic acid, a polymer intermediate, was prepared from levulinic acid (*6*) and used for synthesis of various coatings (*7*). Low-cost levulinate feedstocks resulting from process innovations could revive the levulinate industry and achieve the goal of converting waste to chemical products or intermediates for solvents, surfactants, and resins, including inexpensive binders.

Thermal processing of carbohydrates in nonaqueous solvents containing alcohols offers a direct conversion to levulinate esters that are useful for fuels and chemical intermediates (Figure 1). Fructose was converted to levulinate esters by heating with ethanol and a solid acid (*8*). When the reactions of lignocellulosics such as wood and biomass wastes were conducted in alcohol with strong acid catalysis in an autoclave at temperatures greater than 195°C, the levulinate and formate esters of the alcohol were obtained, along with small amounts of furfural (*9*). A German patent was awarded for a chemical process that decomposes cellulosic waste in alcohols (*10*). In the case of paper wastes and pulps, a stirred precooking stage was required. At lower temperatures, incomplete conversion of cellulose occurred and alkoxymethylfurfural was obtained. The alkoxymethylfurfurals are presumed to be intermediates in the formation of alkyl levulinates and alkyl formates via cleavage of the aldehyde group on the furfural intermediate in a several-step sequence.

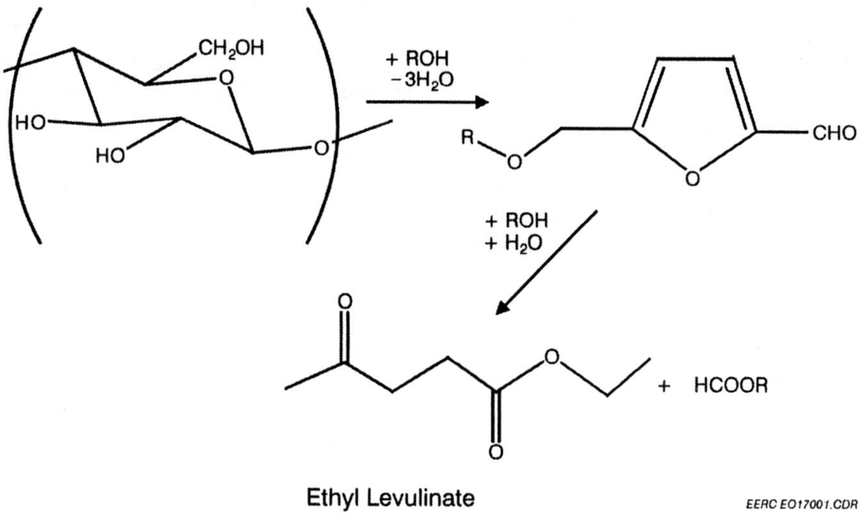

Figure 1. Conversion of cellulose to esters.

Garves reported formation of insoluble humic residue at the higher temperatures (*9*). In our work, the acid-catalyzed reactions of ethanol with seed hulls and with chaff or straw waste at 200°C produced a charcoal product with modest surface areas (*11*). A significant part (20%–30%) of the cellulose is carbonized to charcoal, which is easily removed by filtration. Flash distillation of the excess alcohol and solvent extraction of the levulinate ester leaves a resinous product consisting mainly of the lignin component of the biomass. Thus multiple marketable products are available from processing waste biomass with the acidic alcohols.

Particle Board Waste Processing

Particle board from furniture manufacturing waste presents a unique disposal problem because of the urea–formaldehyde (UF) resin used in binding the wood particles and the polymer-laminated coatings. Many thousands of tons of wood wastes are produced in each furniture plant annually in the form of cut-off pieces and sawdust. The composite materials may form toxic and corrosive gases during combustion, and wastes must be landfilled or burned in an incinerator that safely consumes the resin and plastic materials as well as the wood. Landfilling this waste currently costs millions of dollars per year for a good-sized furniture plant. Thus an environmentally and economically feasible process for converting the composite wood waste to usable products is urgently needed.

As an alternative to landfilling or high-temperature incineration, the acid-catalyzed decomposition in alcohol efficiently and safely converts the composite materials to alkyl levulinate at moderate temperatures. The product composition obtained by heating particle board chips in ethanol with sulfuric acid catalyst for 30 min at 200°C was similar to that obtained from ordinary wood treatment. The charcoal product was removed by filtration; the ethanol, water, and ethyl formate product flash-distilled; and the ester levulinate separated from the resinous product by extraction into diethyl ether. In this case, the resinous products also contained the UF binding resins that were initially present in the waste board. The UF resin-derived components were intimately mixed with or chemically attached to the lignin resin; there was no way to extract and separate the lignin from the UF component.

For this processing, the waste particle board was chipped but not ground to a fine meal. Milling of the particle board is not necessary because the cellulosic material is rapidly decomposed in the acid-catalyzed reaction, which releases the resin into the solvent, resulting in a more economical process. In the Garves patent (*10*), the wood and waste paper were first converted to a pulp and to a fluff, respectively. The moisture content of the as-received particle board was 3 wt%. Waste materials such as this with a low moisture content provide a good substrate for the alcohol reaction, since no drying is required.

Experimental Details

In a typical reaction, a mixture of 16 g of chipped laminated particle board, 150 mL of ethanol, and 3 g of concentrated sulfuric acid was heated at 200°C for 30 minutes in a 300-mL Parr pressure reactor. The mixture was allowed to cool to room temperature. The product mixture was filtered with a Büchner funnel, and the charcoal residue was washed with ethanol and dried to give 5.4 g of charcoal. The filtrate was neutralized with sodium bicarbonate and distilled to recover the excess ethanol and the water, ethyl formate, and trace amounts of diethyl ether produced in the reaction. To extract the levulinate ester, diethyl ether was added to the residue from the distillation, which contained a small amount of ethanol. The insoluble portion of the distillation bottoms gave the resin product (3.3 g). The ether-soluble portion contained ethyl levulinate and a small amount of furfural. After the ether was stripped on a rotary evaporator, the residue (4.5 g) was vacuum-distilled to give ethyl levulinate (3.8 g) and impure furfural (0.4 g).

A larger-scale reaction was conducted with 240 g of chipped laminated particle board in 2.2 l of ethanol in a 1-gallon autoclave at 200°C for 40 min. The heatup and cooldown times for the reaction were much longer, but similar wt% yields of charcoal, esters, and resin were obtained.

Process Variables

A limited matrix of tests was performed to determine effects of process variables, including the type of alcohol used in the processing of particle board. The acid-catalyzed reaction of particle board in ethanol was carried out at several temperatures and reaction times to determine optimum conditions. The concentration of sulfuric acid in ethanol was 2% in all tests. The product yields given in Table I are based on as-received particle board.

Variation of the temperature in the range of 190° to 210°C gave higher levulinate yields at 200° and then no further increase. The longer reaction time (40 min) at 190°C gave superior yields of ethyl levulinate. The weight of the solid char product obtained at 190°C and short reaction times was correspondingly higher, and it is lighter in color and presumably not as carbonized as that obtained at 200°C. The yields of resin were similar for most of the reactions.

Table I. Effect of Temperature and Reaction Time on Ethyl Levulinate Yields

Temperature (°C)	*Reaction Time (min)*	*Ethyl Levulinate Yield (wt%)*	*Solid Yield (wt%)*	*Resin Yield (wt%)*
190	20	11	41	30
190	30	19	44	21
190	40	31	33	21
200	20	14	44	25
200	30	24	34	21
210	30	24	34	21

Yields for several alcohols were determined for the reaction of particle board at 190°C for 40 min reaction time (2% sulfuric acid catalyst). The results for the various alcohols are given in Table II. Methanol and ethanol are essentially equivalent in forming the alkyl levulinate on a weight percent basis. For longer-chain alcohols, the yields drop off substantially with size. In the case of the aminoethanol, methoxyethanol, and hexanol, the solid material is recovered as brown granules and cannot be dried easily. No alkyl levulinate is found in the solvent phase. By comparison with water as the solvent, the charcoal yield was 46 wt% and the levulinic acid was 18% as determined by gas chromatography.

A reaction was performed with plywood sawdust in ethanol at 200°C. This reaction produced a similar yield of ethyl levulinate (23 wt%) and only a little less resin. A comparison of these results with those of Garves (*9*) shows that similar temperatures are required for the various cellulosic materials. Garves obtained higher wt% yields from beech sulfite pulp (30.6 wt%) and lower yields from waste paper fluff (18.5 wt%) for reactions in methanol.

Table II. Formation of Alkyl levulinates in Reactions of Particle Board in Various Alcohols

Alcohol	*Alkyl Levulinate Yield (wt %)*	*Solid Yield (wt%)*	*Resin yield (wt%)*
Methanol	23	36	21
Ethanol	24	34	21
1-Butanol	15	37	25
2-Aminoethanol	0	–	–
2-Methoxyethanol	0	–	–
1-Hexanol	0	–	–

Product Quality

The purity of the ethyl levulinate fractions was determined by gas chromatography. The ether extract contained 85% ethyl levulinate, 10% furfural, and about 2% levulinic acid. The distilled ethyl levulinate cut was generally 97% to 98% ethyl levulinate with 2%–3% levulinic acid. The crude extract should be appropriate for use directly as a fuel oxygenate or for conversion to a fuel such as methyl THF. The distilled products would be suitable for use as a fragrance or for most chemical reactions.

The recovered resin mixture was very tacky and has potential for recycling as a board resin. It was, however, very dark-colored and evolved a slight odor of levulinate ester. It may be used as a fine coal binder for making briquettes or as a waterproof granular fuel. Additional work is being conducted to determine the potential for wood binder recycling, but considerable interest has been shown in the low-cost coal binder applications, and the market could easily consume all the resins that can possibly be produced.

The charcoal is moderately active even without further steam activation. The surface area was determined by the iodine number method (456 mg I_2/g). Steam activation converted the initial charcoal product to a higher-surface-area activated carbon with the iodine number 540 mg I_2/g, but with loss of about half of the carbon. The initial charcoal should be adequate for a number of decolorizing and other aqueous treatments.

Advantages of Cellulose Degradation in Ethanol

Performing the reaction of cellulosic materials in ethanol rather than water offers the advantage of a more direct route to the ester. This provides a convenient one-step reaction to produce a fuel oxygenate additive at minimal cost, compared with two-step reactions involving formation of the acid in the aqueous conditions.

A second advantage of the ethanol process is the ease of product workup into the various product fractions. The charcoal product is filtered from the soluble products; the ethanol solvent, ethyl formate, and product water are readily stripped; and the ethyl ester is easily removed from the resin and distilled without decomposition. In contrast, the purification of levulinic acid is more difficult from a water solution, and consequently the production of levulinic acid is limited *(11)*.

Conversions of Alkyl Levulinates

A major goal for promoting commercialization of the alcohol processing of waste particle board and plywood is the expansion of markets for alkyl levulinates. The major current market for alkyl levulinates is in fragrances, where the ethyl levulinate has a value of about $10/lb. Butyl levulinate is also used but the market is relatively small for both esters.

Both the keto and the carboxylate functional groups of levulinic acid and esters are available for derivatization, and a large number of transformations of either or both groups have been investigated in previous studies *(12–14)*. Levulinate esters of higher alcohols and glycols have been used as plasticizers, especially for vinyl chloride polymers *(15)*.

Transesterification of Methyl and Ethyl Levulinates

Since the yields of the alkyl levulinates were higher with the lower alcohols, methanol, and ethanol, the acid-catalyzed (Fischer) transesterifications of methyl and ethyl levulinate were investigated for the preparation of a number of longer-chain alkyl esters. The reactions were carried out by using a 40%–100% excess of the longer alcohol and 1.7–2 wt% sulfuric acid in the mixture. The methanol or ethanol byproduct was slowly distilled out of the reaction flask through a small distillation column after a 60-min initial reaction time. The higher alcohol was then removed by distillation or vacuum distillation, followed by vacuum distillation of the ester. Overall yields were about 90% of theoretical. The product contained very small amounts of the reacting alcohol. The purities of the product esters determined by gas chromatography are shown in Table III.

Vinyl levulinate has potential for forming copolymers containing keto groups that could help plasticize material. In addition, the keto groups in the copolymers could be further derivatized to useful functionality or to provide cross-linking. The keto group of levulinate can be converted to stable *t*-butyl peroxide derivative or a cyclic peroxide, useful for initiating free radical polymerization.

An attempt to convert methyl levulinate to vinyl levulinate was unsuccessful (Figure 2). Heating methyl ester with vinyl acetate in acidic mercuric acetate gave α-angelica lactone as the major product rather than the vinyl ester. Vinyl levulinate was previously prepared by the mercuric ion-catalyzed reaction of levulinic acid with vinyl acetate (*16*), so this older preparation will be used for future studies. α-Angelica lactone can also be prepared by dehydration of the acid without the use of vinyl acetate (*17*).

Table III. Levulinate Esters Prepared by Transesterification

Starting Ester	*Product Ester*	*BP (°C)*	*Purity*
Ethyl levulinate	chloroethyl levulinate	127–128 (10 mm)	94%
Methyl levulinate	allyl levulinate	106–108 (10 mm)	93%
Ethyl levulinate	decyl levulinate	180–185 (10 mm)	95%
Ethyl levulinate	oleyl levulinate	190–200 (2 mm)	95%

Condensation Reactions

Levulinic acid is known to react with aldehydes at carbon atoms adjacent to both the acid (α) and keto (β and δ) groups to form a mixture of the α-, β-, and δ- alkylidene or arylidene derivatives (*11,12*). With excess formaldehyde, levulinic acid is substituted with five or six hydroxymethyl (methylol) groups, and cyclization to a lactone occurred (*18*). These condensation reactions could be utilized to produce inexpensive resins from the crude ethyl levulinate obtained from waste particle board without distillation. Another reason for studying the reactions of the ester with aldehydes is to determine the potential for similar reactions occurring during the cellulose

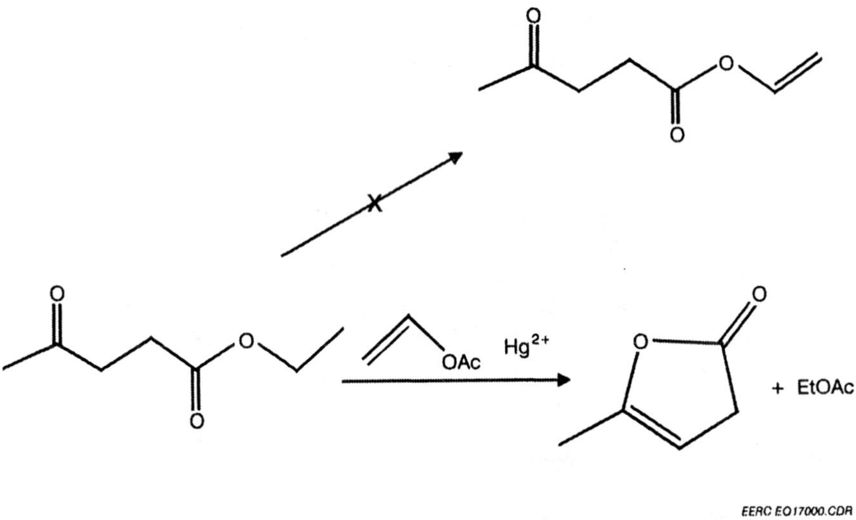

Figure 2. Attempted conversion of ethyl levulinate to vinyl levulinate.

degradation reaction or the subsequent work up of products where aldehydes such as furfural are present.

The reactions of ethyl levulinate with several aldehydes were examined. The reaction of formaldehyde with ethyl levulinate was conducted with dilute sulfuric acid catalyst; the methylol derivatives undoubtedly formed but were not isolated. The reaction with formaldehyde in ethanol solution formed a sticky resin, which was an excellent binder.

The reaction of ethyl levulinate with furfural gave a condensation product, of which the major isomer is presumed to be the δ-furfurylidenelevulinate ester, since the acid-catalyzed reaction of furfural with levulinic acid produces this isomer (*19*). Some of this reaction could occur during the preparation of ethyl levulinate, since furfural and acid catalyst are present, but no furfurylidenelevulinate ester was detected in the ether extract. Thus, if condensation reactions occur, the products stay in the char or the resin products and are likely of higher molecular weight.

In order to selectively obtain the α-substituted condensation products from levulinate esters, it is necessary to protect the ketone group. Ethyl levulinate reacts with one mole of ethylene glycol to form the cyclic ketal. Condensation of the protected ester with benzaldehyde gave the α-benzylidene ester in 73% yield (Figure 3).

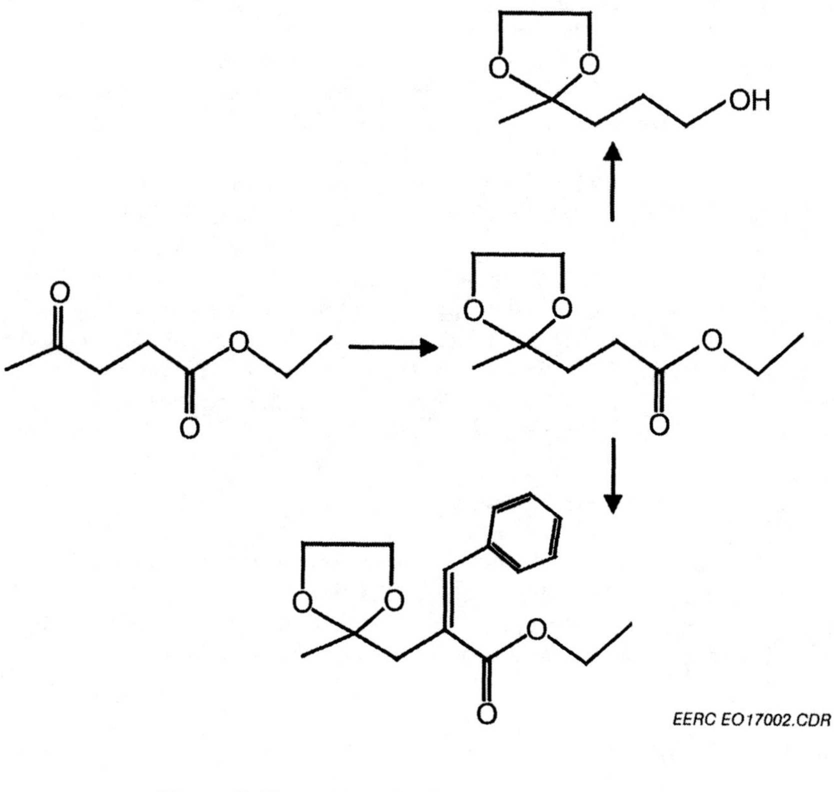

Figure 3. Formation of α-benzylidene derivative.

Resin Intermediates

The acid-catalyzed condensation of ethyl levulinate with urea was investigated. After the mixture was heated, ethanol evolved, so it is presumed that the condensation involved both carbonyl groups of the levulinate ester. The reaction produced a sticky resin that is soluble in alcohol.

A similar reaction of ethyl levulinate was conducted with a UF preparation. The acid-catalyzed reaction incorporated the levulinate into the resin structure, forming an insoluble mass. This reaction does not occur to a large extent during the initial particle board degradation owing to the large volume of ethanol solvent; however, it can cause a problem in the workup of the products after the ethanol has been removed. The reaction explains the failure of attempts to distill ethyl levulinate directly from the levulinate–resin product mixture without prior extraction of the ester from the resin. These attempts resulted in low ester yields and the formation of intractable resin. Thus, by extracting the ester we can avoid the inadvertent condensation reaction with the resin.

Levulinic acid condenses with two moles of phenol to give 4,4-*bis*-(4'-hydroxyphenyl)pentanoic acid (diphenolic acid). This bisphenol intermediate was used to prepare several types of resins, including phenolic–formaldehyde, epoxy, polycarbonate, polyester, and polyurethane types (*7*). Ethyl levulinate undergoes a similar reaction with phenol to give the diphenolate ester (Figure 4). Using a modification of the Bader procedure (*6*), the ethyl 4,4-bis-(4'-hydroxyphenyl)pentanoate (ethyl diphenolate) was obtained in 67% of theoretical yield based on starting ethyl levulinate ester. The published reaction with levulinic acid gave 60% yield, based on starting levulinic acid.

Further conversions to resins were performed with the diphenolate ester intermediate. Polymeric linkages can be effected through the two phenolic groups in typical bisphenolic syntheses, or the carboxylate can be used as a third linkage in cross-linked resins. Alternatively, the ester functionality can be easily transformed to amide and other groups with interesting binding properties. An advantage in this type of polymer is the extension of the ester or amide functionality away from the polymer backbone. The properties of the carboxy-modified resins are currently being investigated.

Particle Board Waste Process Design

A schematic for waste particle board processing is shown in Figure 5. This process avoids heating the resin in the presence of ethyl levulinate, which generates a highly insoluble resin product. Instead, the ethyl levulinate is extracted from the resin and subsequently vacuum-distilled to purify it.

Plant Design

A preliminary plant design and cost analysis for the particle board processing was performed (*20*). This projection used extensive distillation

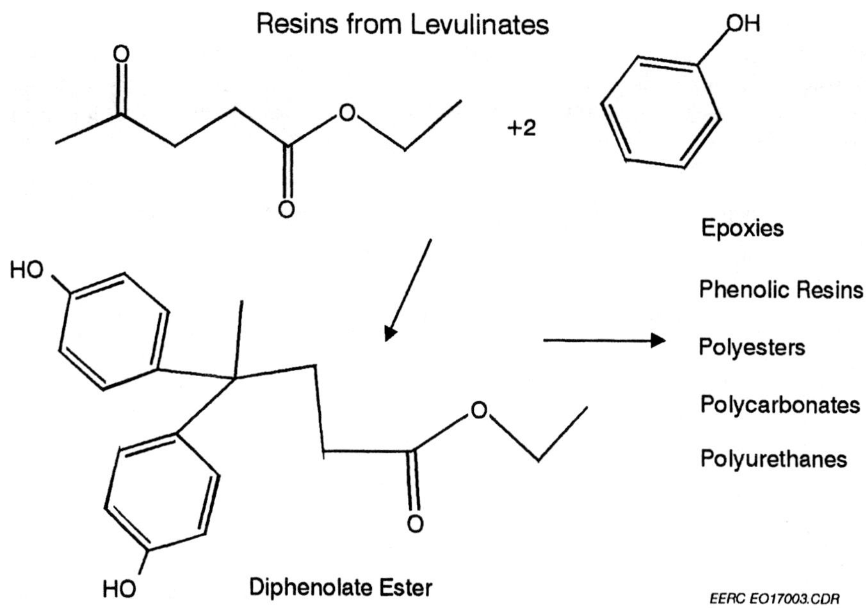

Figure 4. Formation of diphenolate ester.

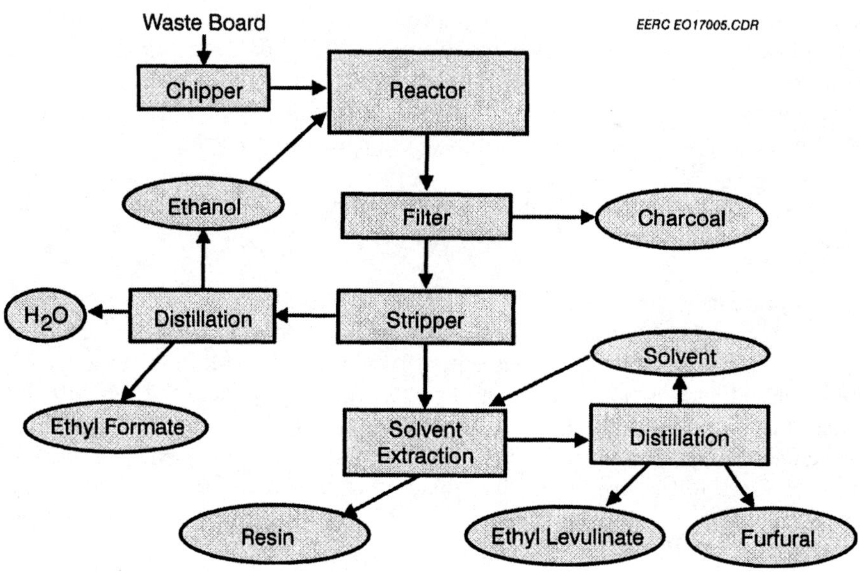

Figure 5. Particle board waste process.

equipment to produce pure ester products from waste particle board input at 33,000 tons/year. Product output is shown in Table IV. Assuming that the value for the ethyl levulinate is \$0.15/lb used as fuel additive, the analysis showed that the after-tax internal rate of return (IRR) would be 3.4%, which is not enough to justify building a plant for this market. If, however, the value of ethyl levulinate as a solvent and chemical intermediate could be increased to \$0.60/lb, the sensitivity analysis showed that the IRR would be an acceptable 20%.

Table IV. Process Products for 33,000-ton Plant

Product	*Yearly Production (million lb)*	*Estimated Market Value (\$/lb)*
Ethyl Levulinate	15	0.15–60
Ethyl formate	7	0.50
Furfural	1	0.70
Charcoal	10–22	0.25
Resin	15	0.10

Thus, more development work needs to be done to understand the value of the reaction products (esters, charcoal, and recovered resin) in various chemical markets and determine if fuel use is the only option for developing the large market needed for the ethyl levulinate produced. If the fuel oxygenate additive market is the only large market option, less distillation processing would be required, and revision of the process and plant design is needed. The production cost could be much lower, since the cost of the distillation columns was 65% of the purchased equipment cost in the plant design described above.

Summary

In summary, conversion of waste particle board and plywood directly to levulinate and formate esters utilizes waste that represents a disposal problem in a process that generates easily purified products with minimal wastewater. Commercialization of this process depends on market values for levulinate esters that are typical of chemical intermediates for resins and coatings. New challenges are presented in finding ester reactions and subsequent purification steps that lead to low-cost derivatives that will be useful for these industries. The economics may also be favorable for producing ester fuel additives with no purification required.

References

1. Dunlap, A.P. U.S. Patent 2,813,900, 1957.
2. Fitzpatrick, S. U.S. Patents 4,897,497, 1990; 5,608,105, 1997.
3. Parkinson, G., Ed. *Chem. Eng.* 1998, *Sept., 25.*
4. Eckert, G.W. U.S. Patent 3,074,787, 1963.
5. Thomas , J.J.; Barile, R.G. *En. Biomass Wastes* **1984**, *8*, 1461.
6. Bader, A.R.; Kontowicz, A.D. *J. Amer. Chem. Soc.* **1954**, *76*, 4465.
7. Bader, A.R. U.S. Patent 2,933,472, 1960.
8. Brown, D.W.; Floyd, A.J.; Kinsman, R.G.; Roshan-Ali, Y. *J. Chem. Tech. Biotechnol.* **1982**, *32*, 920.
9. Garves, K. *J. Wood Chem. Technol.* **1988**, *8*, 121.
10. Garves, K. German Patent DE 3621517 A1, 1988.
11. Kitano, M.; Tanimoto, F.; Okabayashi, M. *Chem. Economy Eng. Rev.* **1975**, *7*, 25.
12. Quaker Oats Co. *Levulinic Acid*; Bulletin 301A.
13. Shilling, W.L. *Tappi* **1965**, *48*, 105A.
14. Leonard, R.H. *Ind. Eng. Chem.* **1956**, *48*, 1331.
15. Emerson, W.S.; Langley, R.I. U.S. Patent 2,581,008, 1952.
16. Kohlhammer, K.; Huster, W.; Deinhammer, W. U.S. Patent 5,278,225, 1994.
17. Wolff, L. *Ann.* **1885**, *229*, 249.
18. Olsen, S. *Acta Chem. Scand.* **1955**, *9*, 101.
19. Urban, R.S. U.S. Patent 2,688,621, 1954.
20. Mann, M. Personal communication, 1999.

Chapter 6

Carbohydrate Diacids: Potential as Commercial Chemicals and Hydrophobic Polyamide Precursors

D. E. Kiely

Shafizadeh Rocky Mountain Center for Wood and Carbohydrate Chemistry, University of Montana, Missoula, MT 59812

Carbohydrate diacids (aldaric acids) are derived from oxidation of the terminal carbons of simple aldoses. Several aldaric acids are suggested as being reasonable candidates for commercial development, based on potential applications, and availability and cost of the individual simple sugar precursors. Given that D-glucose, produced on a massive scale worldwide by starch hydrolysis, is by far the most available and inexpensive aldose, the corresponding aldaric acid, D-glucaric acid, would appear to be the most attractive aldaric acid for commercial development. This report will focus on several potentially important aldaric acids, consider some of their applications as standalone chemicals, and describe how they can be used as diacid monomers for the preparation of assorted polyamides, labeled polyhydroxypolyamides (PHPAs) or Hydroxylated Nylons.

The term "carbohydrate diacids" as used in this report represents that class of compounds know as aldaric acids, i.e., simple diacids derived from oxidation of the terminal carbons of aldoses. As a class of compounds aldaric acids have been known for more than a century and some of these diacids played an important role in Emil Fischer's assignment of the relative configurations of naturally occurring D-aldoses *(1)*. Surprisingly, these molecules have seen very little chemical application over the past century but their time may have arrived to be considered as valuable and versatile synthons for preparation of specialty chemicals and polyamides. The material in this report presents an overview of how, in recent years, some of these compounds have been synthetically employed in the author's laboratories, and in those of other investigators. It is hoped that the examples illustrated here will stimulate further development of these small but synthetically versatile molecules.

Aldaric Acid Preparation

Historically, aldaric acids have been prepared principally by nitric acid oxidation of aldoses *(2)*, although catalytic oxidation of aldoses has had some success *(3)*. The nitric acid oxidation method is remarkably selective and in general yields the corresponding aldaric acid (in one form or another) as the major product. Given the long history of this reaction, it is a wonder that aldaric acids haven't been produced in bulk quantities for years and become available and inexpensive compounds. Quite the contrary, only a limited number of aldaric acids can be purchased from chemical suppliers and those are quite expensive. In Figure 1 are structures of aldaric acids that might be considered as potential bulk chemicals, a consideration based primarily on the availability and cost of the starting sugar precursors.

Among the acids shown, only four carbon L-tartaric acid (**1**), produced as a byproduct of the wine industry, is at present a relatively important commercial aldaric acid. However, five carbon L-arabinaric acid (**2**) and *meso*-xylaric acid (**3**), while not available from chemical suppliers as far as this author is aware, could be produced by appropriate oxidation of L-arabinose (from hydrolysis of the plant polysaccharide, arabinogalactan) and D-xylose (wood sugar), a hydrolysis product of woody plant hemicellulosic xylans. In contrast the three six carbon aldaric acids shown, while costly, are available from a number of laboratory chemical suppliers. Crystalline *meso*-galactaric (mucic) acid (**4**) is synthetically the most well known aldaric acid as its preparation by nitric acid oxidation of lactose represents a classical procedure in synthetic organic chemistry *(4)*. D-Mannaric acid ***(5)*** is typically prepared by nitric acid oxidation of D-mannose *(5)*, and is sold as its crystalline 1,4:6,3-dilactone. The last and potentially most commercially attractive aldaric acid is D-glucaric acid (**6**), a direct nitric acid *(2)* or platinum catalyzed *(3)* oxidation product of D-glucose (dextrose). In terms of a large scale bulk chemical, D-glucaric acid is the only acid among this group that offers the opportunity for economy of scale. Precursor D-glucose is inexpensive and available as a major commodity chemical from starch hydrolysis, and in particular corn starch hydrolysis.

D-Glucaric acid, while typically represented in a Fischer projection formula as an acyclic product, is not isolated in this form but is isolated and easy to handle as crystalline 1,4- and 6,3- monolactones, **6a** and **6b**, respectively, or the 1,4:6,3-dilactone (**6c**). Several convenient salt forms of acyclic D-glucaric acid can also be prepared and these include monopotassium (**6d**) - *(6)*, dipotassium (**6e**) - *(7)*, sodium potassium (**6f**) - *(7)*, and calcium D-glucarate (**6g**) - *(6)*. In aqueous solution glucaric acid is present as an equilibrium mixture of principally the acyclic structure, and the 1,4- and 6-3-monolactones *(8,9)*.

L-Tartaric Acid

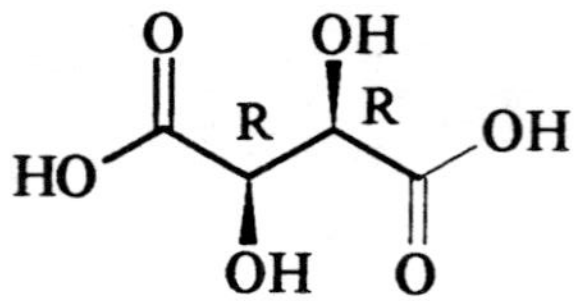

1

L - Arabinaric Acid

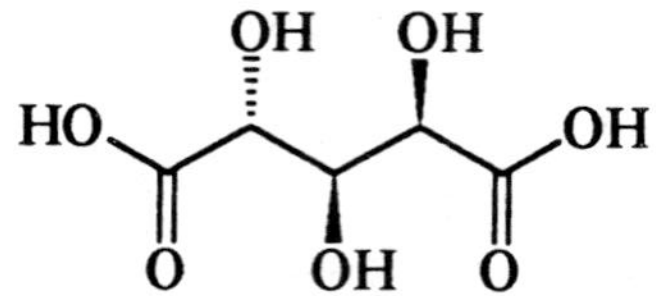

2

***meso*-Xylaric Acid**

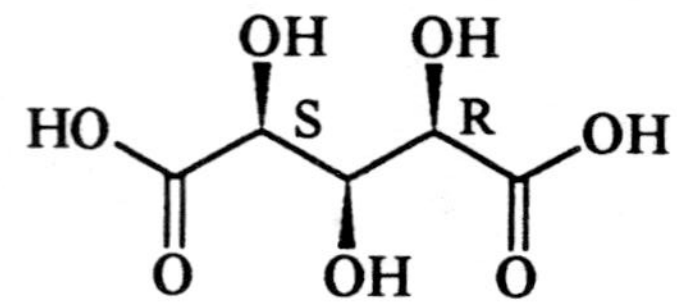

3

***meso*-Galactaric Acid**

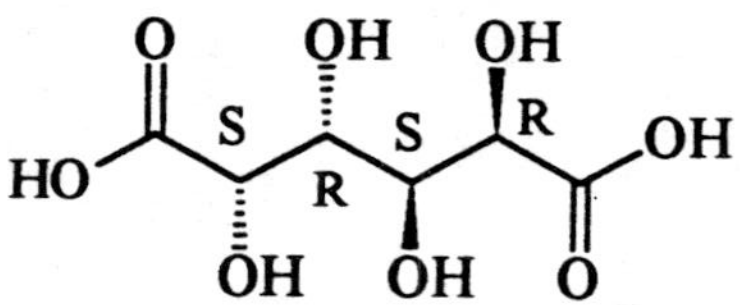

4

D-Mannaric Acid

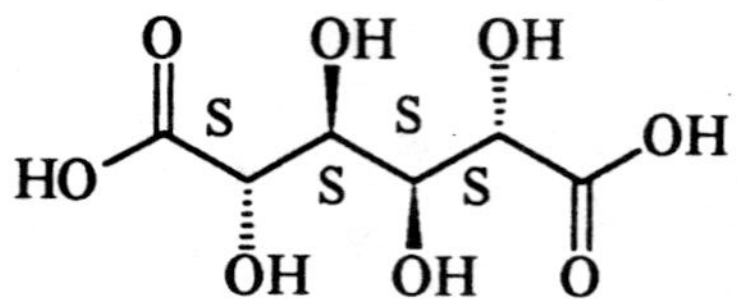

5

D-Glucaric Acid

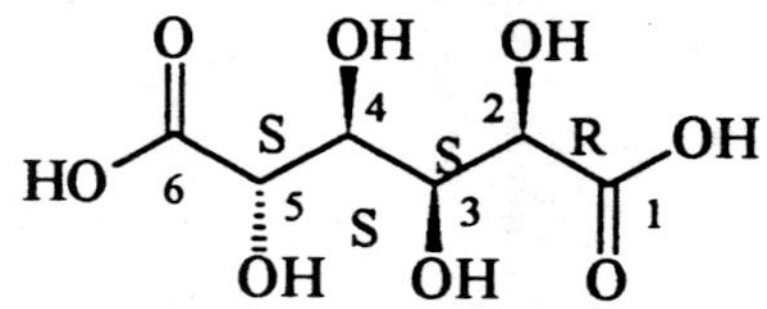

3

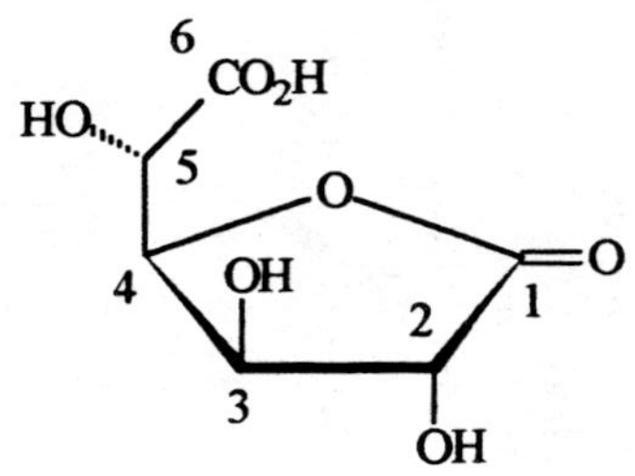

6a

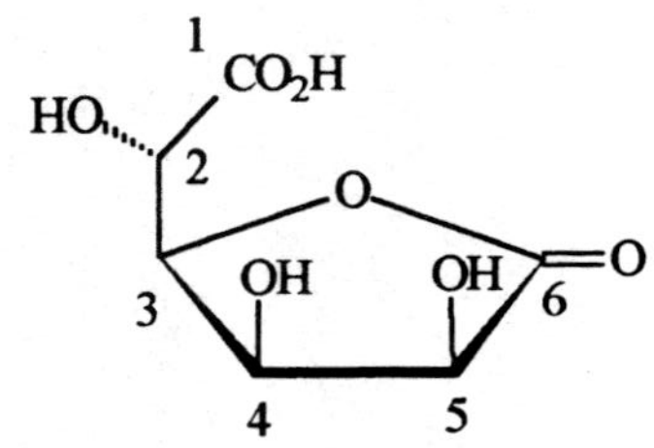

6b

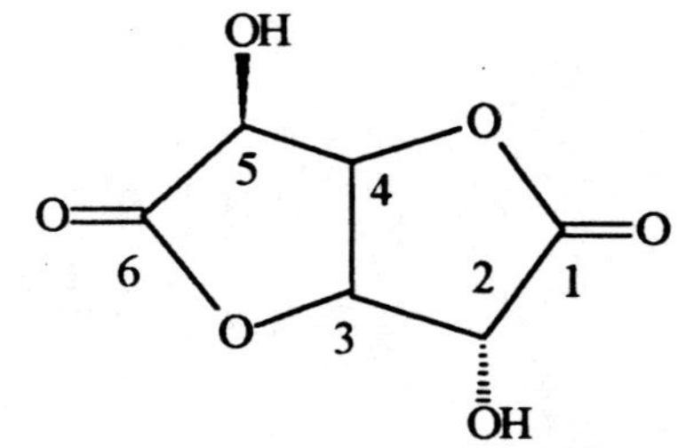

6c

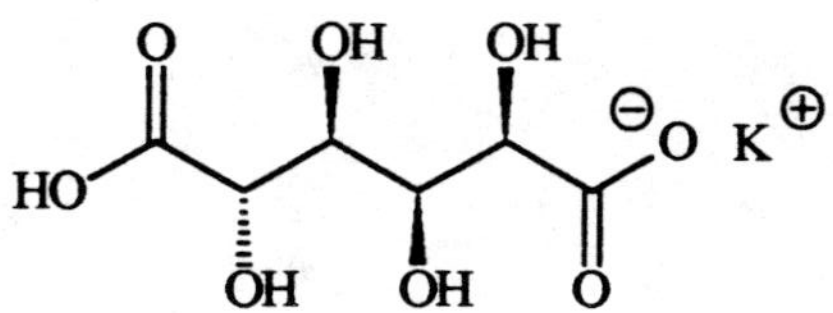

6d

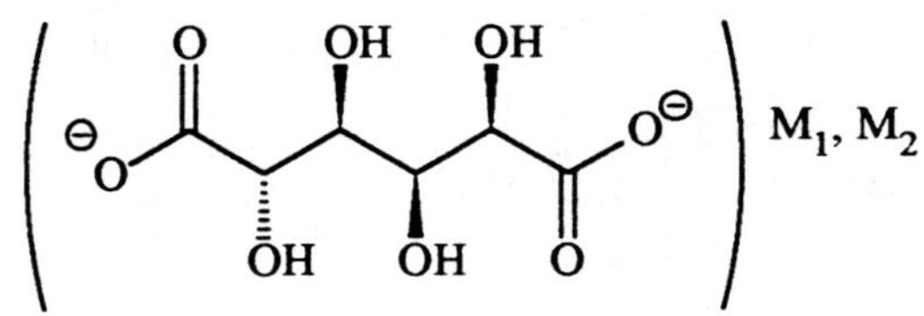

6e - $M_1 = M_2 = K^{\oplus}$
6f - $M_1 = Na^{\oplus}, M_2 = K^{\oplus}$
6g - $M_1 = Ca^{\oplus\oplus}$

Figure 1. Some important aldaric acids and salts

Aldaric Acids as Synthons

Polyhydroxypolyamide Synthesis

Our current major interest in aldaric acids is as monomers for polycondensation reactions with diamines to yield a broad class of polymers called polyhydroxypolyamides (PHPAs) or Hydroxylated Nylons *(10)*. The original reports of examples of this class of polyamides were from the laboratory of N. Ogata and coworkers in Japan, beginning in the 1970s *(11)*. More recent reports on these polymers have come from the laboratories of Hashimoto and coworkers *(12)* and those of author *(13)*. Our general preparation of these polyamides, exemplified with a generic hexaric acid, is outlined in Figure 2.

Typically, condensations to polyamide (**II**), are carried out between an activated, i.e., esterified (diester, ester/lactone, dilactone) form of the aldaric acid (**I**), and a primary diamine in a polar protic solvent such as methanol at room temperature. Under those conditions any starting single "ester" form of a five or six carbon aldaric acid is rapidly converted to an equilibrium mixture of acyclic diester and ester/lactone forms. Dilactone, if formed at all, is present in only small amounts. The starting monomer "esterified" form(s) of the aldaric acid differ from one acid to an other. For example, the monomers of choice for L-tartaric acid (**1**) and *meso*-galactaric acid (**4**) are simple diesters, while for D-glucaric acid suitable monomers are ester/lactones, dilactone, or even the entire alcohol esterified mixture *(10)*. Xylaric acid is also conveniently polymerized as its esterification mixture while the D-mannaric acid monomer is the 1,4:6,3-dilactone *(13a)*.

Given the ease of carrying out these polymerizations, we have prepared a number of polyamides in order to survey how reaction conditions and change in diacid and/or diamine monomers influence the reaction itself and the properties of the polymers obtained. We recently described a number of PHPAs from D-glucaric, *meso*-xylaric, *meso*-galactaric and L-mannaric acid with an assortment of diamines, and our studies indicated several trends *(13a)*. Some typical diamines employed in the polymerizations are shown in Figure 3. Polyamides from conformationally "bent" D-glucaric and *meso*-xylaric acids tend to be more water soluble and have lower melting points then those derived from conformationally "extended" *meso*-galactaric and D-mannaric acids. As the alkylenediamine monomer units become shorter, polyamide water solubility increases. Whereas the poly(alkylene D-glucaramides) from ethylenediamine and tetramethylenediamine are water soluble, those based on hexamethylendiamine to dodecamethylenediamine become increasingly less water soluble. Interestingly, all of these latter polymers have similar melting points (ca. 190 °C) in spite of their range of water solubility. Thus, it is clear that variable degrees of both hydrogen bonding (aldaryl unit) and van der Waals attractive forces (alkylene unit) impact the properties of the polymers. It was further observed that polyamides derived from branched diamines and/or those containing a heteroatom in the backbone had enhanced polymer water solubility,

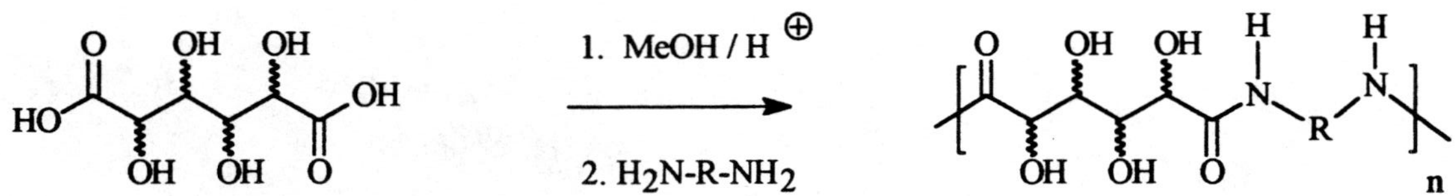

I

II

Figure 2. General procedure for converting an aldaric acid, illustrated with a generic hexaric acid **I**, to a polyhydroxypolyamide (PHPA), **II.**

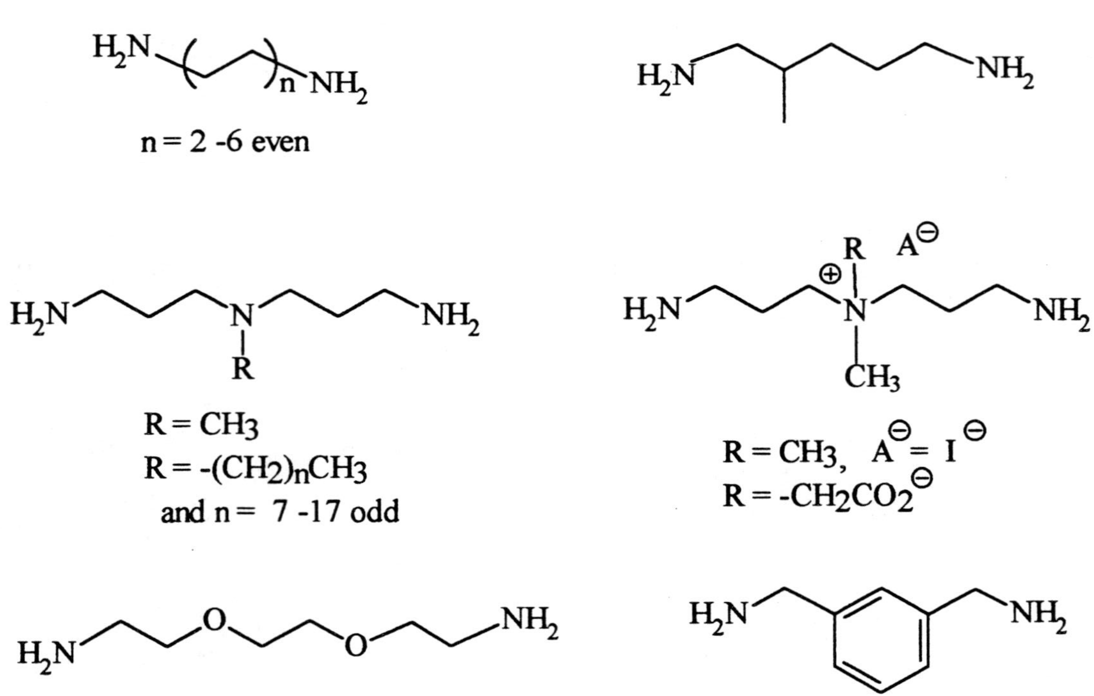

Figure3. Some typical diamine repeating monomer units in the polyhydroxypolyamides

film forming properties, and adhesive characteristics compared to the poly(alkylene aldaramides) *(13b,13c)*. Three examples of polyamides, with different aldaric acid monomer units, that showed relatively good film forming and adhesive properties are shown in Figure 4. In general, polymers from diamines with heteroatoms in the diamine backbone or branched diamines had superior film forming and adhesive properties than those polyamides with high alkylenediamine monomer content.

Several aspects of this research are particularly exciting. The condensation polymerization method allows for preparation of a broad spectrum of polymers with different primary structures based simply on using different diacid and diamine monomers. As a result, by choice of the acid and diamine monomers, one can alter, in somewhat predictive fashion, such polymer properties as melting point, solvent solubility, crystallinity, film and adhesive characteristics and undoubtedly other polymer properties that remain to be studied. Furthermore, the condensations require no aldaric acid hydroxyl group protection/deprotection, a very important cost consideration in the potential production of any of these materials.

This chemistry also provides the option to prepare stereoisomeric polyamides, again with different properties and potentially different uses. For example, condensing a common diamine monomer with D-mannaric, *meso*-galactaric and D-glucaric acid generates diastereoisomeric polyamides. Whereas the poly(D- mannaramides) are repeating *head-tail* polymers (with C-1 and C-6 of the aldaryl unit representing the head and tail ends, respectively), *(12b,13a)*, the poly(galactaramides) have a random distribution of the head and tail ends of the repeating galactaryl unit, as do the poly(D-glucaramides) prepared as described. However, stereoregular repeating *head, tail*-poly(D-glucaramides) *(14)* and alternating *head, tail - tail, head*-poly(D-glucaramides), *(15)*, can be prepared, using modified procedures, but also without hydroxyl protection/deprotection.

Small molecule synthesis

Figure 5 illustrates the use of *meso*-xylaric acid as a precursor for the preparation of a number of synthetic six-membered carbocylic *(16)* and nitrogen heterocyclic products *(17)*. The routes to the carbocylic products shown were similar and involved first a two carbon extension of tri-*O*-acetylxylaryl dichloride (**7**) with diazomethane to the corresponding bis(diazomethyl)ketone which was readily converted to bis(halo- or azidomethyl)diketones (**8**). The latter undergo efficient and convenient cyclization to branched acetylated cycloses which in turn are easily converted to the branched inositols or six-membered unsaturated carbocycles shown. The diamide **10**, derived from the same xylaryl dichloride, and the diamide **11**, from ribaric acid, both served as direct precursors for the nitrogen heterocycle illustrated.

The six membered carbocyclic compounds shown in Figure 5 were prepared by us in a study designed to prepare new branched cyclitols. Cyclitols, both

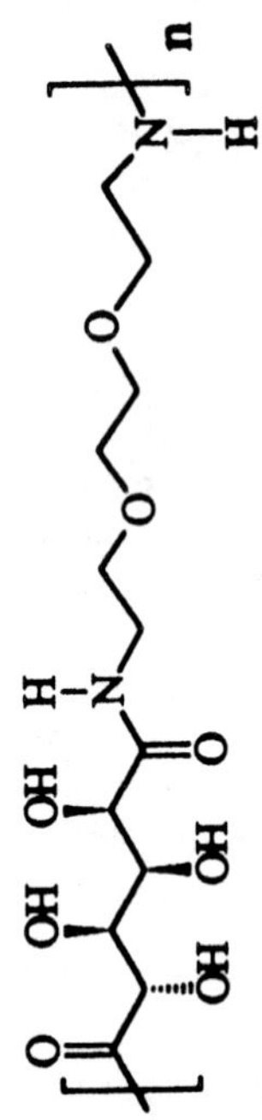

glucaric

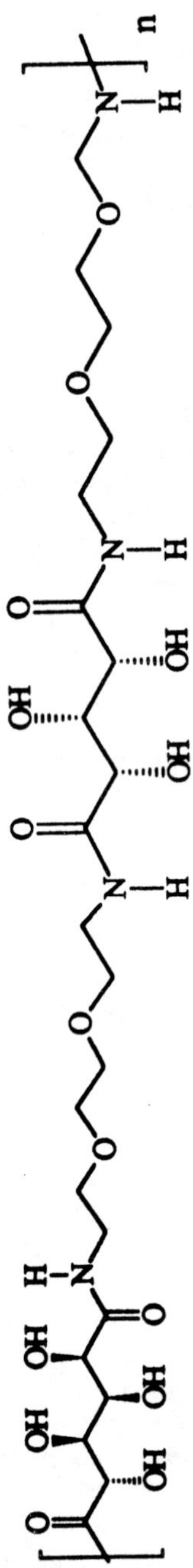

glucaric

xylaric

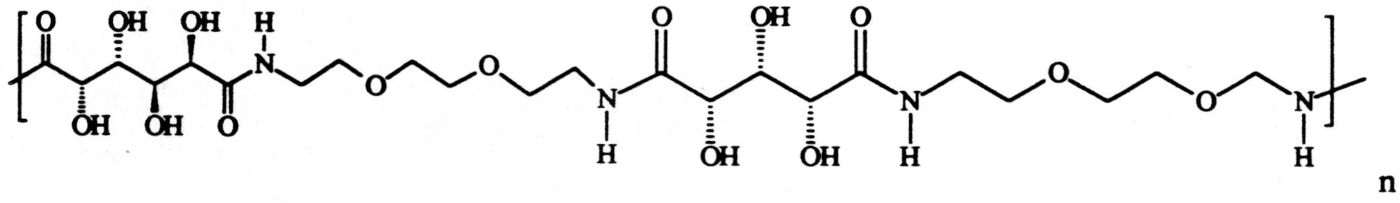

galactaric

xylaric

Figure 4. Some polyhydroxypolyamides with adhesive and film forming propertiesderived from 2,6-dioxaoctane-1,8-diamine and different aldaric acids.

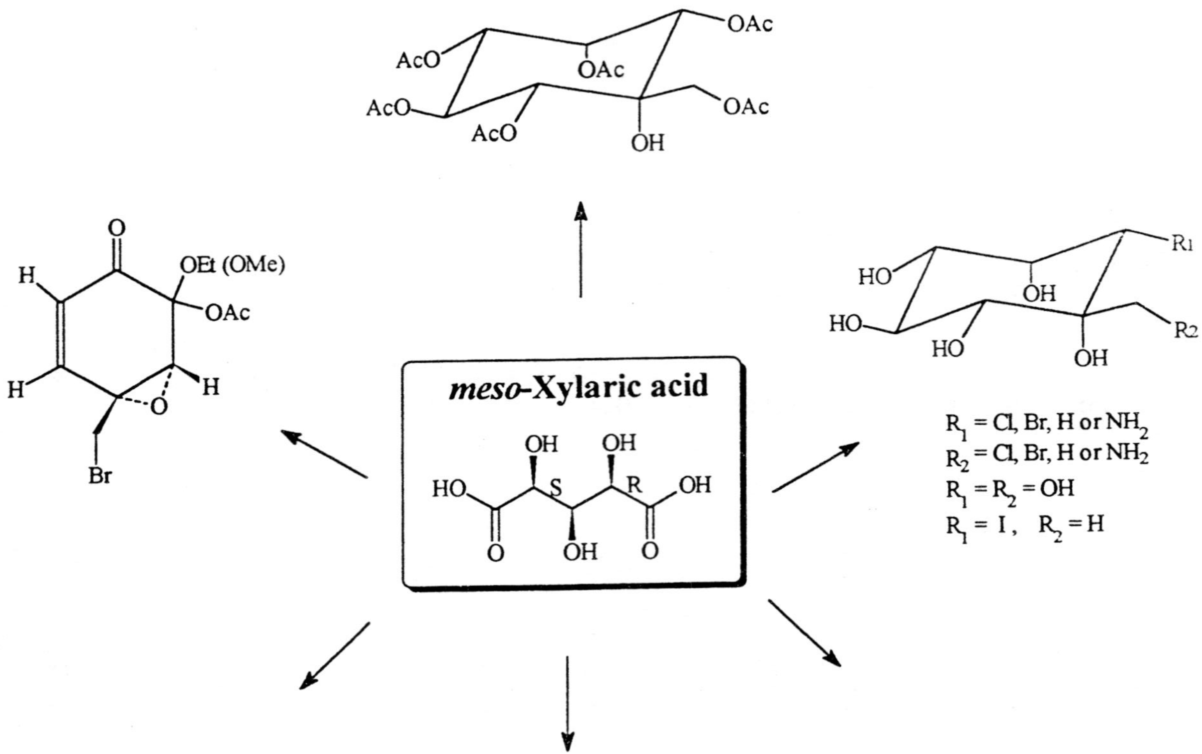

meso-Xylaric acid
OH
OH
HO
S
R
OH
O
OH
O
AcO
OAc
OAc
AcO
AcO
OAc
OH
O
H
OEt (OMe)
OAc
H
H
O
Br
HO
OH
R1
HO
R2
HO
OH
R1 = Cl, Br, H or NH2
R2 = Cl, Br, H or NH2
R1 = R2 = OH
R1 = I, R2 = H

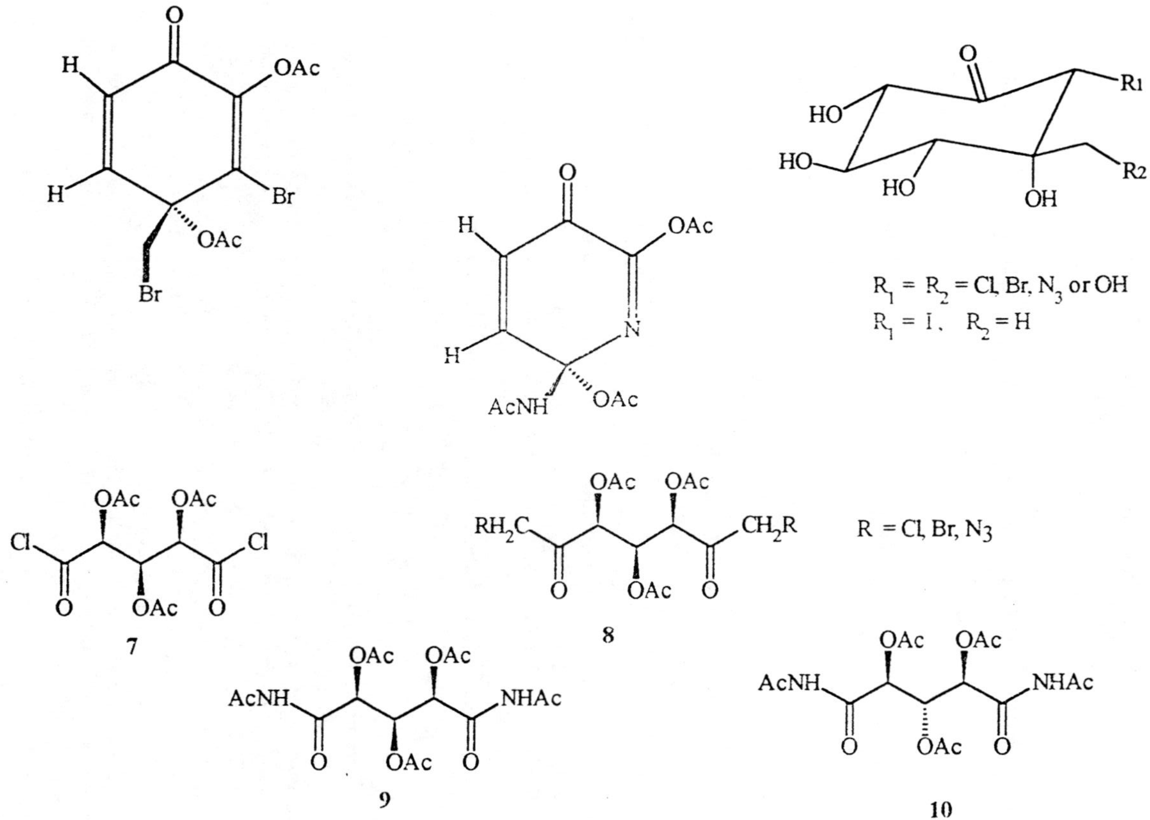

Figure 5. Some branched cycloses, cyclitols and alpha, beta-unsaturated cyclic ketones derived from xylaric acid; some acyclic *xylo* (**8** - **9**) and *ribo* (**10**) ring precursors.

branched and unbranched, were of interest because of their biosynthetic origin from carbohydrates, and as components of aminocylitol antibiotics. The novel cyclitols, unsaturated carbocycles, and unsaturated heterocycle shown were all prepared from xylaric acid, a molecule that has still seen very little use but clearly can be considered as a synthon for preparing small, multi-functional molecules of biological interest. In short, the synthetic value of aldaric acids in organic chemistry and medicinal chemistry is largely unexplored.

Applications of Aldaric Acids and Polyhydroxypolyamides

Glucaric acid applications

Although applications for aldaric acids have not been seriously tapped to date, these molecules clearly offer opportunities as "standalone" chemicals and as chemical precursors for manufacture of other materials, both small molecules and polymers. The applications noted here are those of D-glucaric acid, but in general might also apply to other structurally related aldaric acids. Ultimately, the long range potential economic value of glucaric acid can only be realized when there is at least a moderate supply of less expensive acid available for further applications testing.

Health related applications - D-Glucaric acid is a human metabolite and is generally regarded as a safe compound for human use, making it an attractive compound for health related applications. Currently, the most rapidly growing commercial use of glucaric acid is in the neutraceuticals (diet supplements) area where calcium D-glucarate is being marketed as a nontoxic natural compound for adjuvant cancer prevention, *(18)* and prevention and treatment of hypercholesterolemia *(19)*. Glucaric acid, and other alpha hydroxy acids such as lactic acid, have use in various skin conditioners and skin healing applications *(20)*, skin aging prevention products *(21)*, antibacterial and antiviral compositions *(22)*, and as a hair growth stimulant component *(23)*. It is additionally considered advantageous in the manufacture of a calcium fortified yogurt *(24)*, and calcium fortified powdered milk *(25)*.

Industrial process applications - The following list of applications is taken largely from the patent literature and illustrates the range of applications identified for D-glucaric and/or other aldaric acids: electroplating processes *(26)*, epoxy resin based powder coatings *(27)*, applications in sugar cane *(28)* and sugar beet processing *(29)*, as a cement additive *(30)*, in bleaching textile formulations *(31)*, as a dispersing and milling aid for inorganic fillers and pigments for production of lacquers, dyes, paper, etc. *(32)*, in ceramic materials with dispersed titanium for corrosion resistance *(33)*, and for the manufacture of some antifogging plastics *(34)*.

Other Applications - Other reported applications include glucaric as a component of detergent formulations (assorted patents) *(35-37)*, shampoos *(38)*, calcium and magnesium containing fertilizers *(39)*, fillings (calcium salt) for ceramics and other porous materials *(40)*, flocculating agents for water clarification *(41)*, as an additive in the manufacture of noodles *(30)*, and as a calcium chelating agent for soil stabilization *(42)*.

A number of the applications also list L-tartaric acid and D-glucaric acid as having comparable properties, both being hydroxy acids and chelators. When inexpensive glucaric acid becomes commercially available on a significant scale, it has applications waiting for a supply and could overtake uses now dominated by tartaric acid. Additional applications unique to glucaric acid will also benefit from a dependable source.

Polyhydroxypolyamides Applications

Opportunities for the polyamides described above, like that of glucaric acid itself, have yet to be realized, and will depend in part on a supply of inexpensive glucaric acid. However, any successful commercial ventures with these polymers are likely to take advantage of one or more factors concerning their preparation and properties. A multitude of these polyamides are possible given their ease of preparation, a method that requires no hydroxyl protection/deprotection. Polymer properties such as hydrophobic/hydrophilic balance, film forming and adhesive properties, water solubility/insolubility, orientation of the chiral D-glucaryl unit in the polymer, charge on the polymer, etc., can be manipulated by the choice of glucaric acid - diamine(s) combination, and whether a procedure to prepare random or stereoregular polyamides is used. Furthermore, given the apparent biodegradability of the polymers *(43)*, applications that take advantage of this property also become important. As biodegradable materials, they have very different properties from the more conventional biodegradable polyester materials, such as poly(lactic acid). The polyamides might not compete with the polyesters in some applications, but the polar properties of the polyhydroxypolyamides might make them better candidates than polyesters for other biodegradable materials applications. Given what we know at present about these polyamides the following potential applications are reasonable: biodegradable adhesives *(44)*, specialty time release fertilizers *(43)*, industrial chemicals for the textile and paper industries, water treatment chemicals, detergent components, hydrogel components, composite material components, and specialty film and fiber materials.

The above are optimistic applications targets for the materials, and while the potential for commercial success has yet to be determined, the challenges associated with commercial development of biodegradable materials are significant and well known. Poly(lactic acid) (PLA) was studied for about forty years before it was announced by Dow - Cargill in January of 2000 that the first large scale production facility for this biodegradable polyester derived from corn starch is now scheduled to be built *(45)*.

Time will tell whether D-glucaric and other aldaric acids will, like L-lactic acid, become important chemical building blocks from renewable agricultural resources.

Potential Commercial Markets - Aldaric acids, and in particular D-glucaric acid, represent a class of molecules whose direct use potential and synthetic potential have yet to be realized. Thus one can only speculate on the market potential they may have. Their monetary value will depend on to what extent they become commercially available and at what cost, and how many applications become economical. However, these acids are derived from renewable agriculture resources (in particular starch), and the growing trend toward adding value to ag materials will sooner or later drive their production and applications development. When that occurs we will know how much they are actually worth.

References

1. Fischer, E. *Ber.* **1891**, *24*, 1836, 2683; see also Hudson, C. S. *Advan. Carbohydr. Chem.* **1948**, *3*, 1.
2. Mehltretter, C. L.; Rist, C. E. *J. Agr. Food Chem.* **1953**, *1*,779.
3. Mehltretter, C. L.; Rist, C. E.; Alexander, B. A. U. S. Patent 2,474,089, 1949.
4. Lewis, B. A.; Smith, F.; Stephens, A. M. *Methods Carbohydr. Chem.* **1963**, *2*, 38.
5. Haworth. W. N.; Hesslop, W. N. D.; Salt, E.; Smith, D. *J. Chem. Soc.* **1944**, 217.
6. Mehltretter, C. L. *Methods Carbohydr. Chem.* **1963**, *2*, 46.
7. Styron, S. D.; French, A. D.; Friedrich, J. D., Lake, C. H.; Kiely, D. E. submitted for publication in *J. Carbohydr. Chem.*
8. Hirasaka, Y.; Umemoto, K.; Sukegawa, M.; Matsunaga, I. *Chem. Pharm. Bull.* **1965**, *13*, 677.
9. Horton, D.; Walaszek, Z. *Carbohydr. Res.***1982**, *105*, 95.
10. a) Kiely, D. E.; Chen, L.; Lin, T-H *J. Amer. Chem. Soc.* **1994**, *116*, 571. b) Kiely, D. E.; Lin, T-H U.S. Patent 4,833,230, **1989**.
11. a) Ogata, N.; Hosoda, Y. *J. Polym. Sci. Polym. Lett. Ed.* **1974**, *12*, 355. b) Ogata, N.; Sanui, K.; Iijima, K. *J. Polym. Sci. Polym. Chem. Ed.* **1973**, *11*, 1095. c) Ogata, N.; Hosada, Y. *J. Polym. Sci. Polym. Chem. Ed.* **1975**, *13*, 1793. d) Ogata, N.; Sanui, K. *Polym. Sci. Polym. Chem. Ed.* **1977**, *15*, 1523.
12. a) Hashimoto, K; Okada, M.; Honjou, N. *Makromol. Chem., Rapid Comm.* **1960**, *11*, 393. b) Hashimoto, K.; Wibullucksanakul, S.; Mausuura, M.; Okada, M. *J. Polym. Sci., Polym. Chem. Ed.* **1993**, *31*, 3141.

13. a) Kiely, D. E.; Chen, L.; Lin, T-H. *J. Polym. Sci. Polym Chem. Ed.* **2000**, *38*, 594. b) Morton, D. W.; Kiely, D. E.; *J. Polym. Sci. Polym Chem. Ed.* **2000**, *38*, 604. c) Morton, D. W.; Kiely, D. E., *J. Applied Polym. Sci.* in press.
14. Chen, L.; Kiely, D. E.; *J. Org. Chem.* **1996**, *61*, 35294.
15. Styron, S. D., *Ph.D. Dissertation* **1999**, University of Alabama at Birmingham.
16. a) Cantrell, C. C.; Kiely, D. E.; Abruscato, G. J.; Riordan, J. M. *J. Org. Chem.* **1977**, *42*, 3562. b) Kiely, D. E.; Cantrell, C. C.; Riordan, J. M.; Abruscato, G. J. *J. Carbohydr. Chem.* **1982**, *1*, 49. c) Riordan, J. M.; Kiely, D. E. *Carbohydr. Res.* **1980**, *82*, 303.
17. Kiely, D. E.; Semk-Gray, K.; Riordan, J. M. *J. Carbohydr. Chem.* **1982**, *1*, 191.
18. Walaszek, Z.; Szemraj, J.; Narog, M.; Adams, A. K.; Kilgore; J.; Sherman, U.; Hanausek, M. *Cancer Detect. Prev.* **1997**, *21*, 178.
19. Walaszek, Z.; Slaga, T.; Hanausek, M. PCT Int. Appl., *Chem. Abstr. 116*, 143875p.
20. Gans, E. H.; Shacknai, J. PCT Int. Appl., *Chem. Abstr. 127*, 9127.
21. Yu, J. R.; Van Scott, E. J. Eur. Patent., 92-105805, **1992**, *Chem. Abstr. 126*, 135447.
22. Wolf, F.; Juestel, C.; Schreiber, J.; Klier, M. Eur. Patent 775,486, **1997**, *Chem. Abstr. 114*, 149925.
23. Davis, M. A. F.; Gibson, W. T. Eur. Patent 375,388, **1990**.
24. Hansen, P. M. T.; Fligner, K. U.S. Patent 5,449,523, **1995**.
25. Korte, D. D.; Hansen, P. M. T.; Fligner, K. Eur. Pat. Appl., *Chem. Abstr. 120*, 243040.
26. Martyak, M. N.; Monzyk, B. F.; Chien, H. H. U.S. Patent 5,306,336, **1994**, *Chem. Abstr. 121,* 89355.
27. Weiss, J. V. Eur. Canadian Patent 785241, **1997**, *Chem. Abstr.127*, 163201.
28. Carrazana Ruiz, L.; Morffi Suarez, E. *Cent. Azucar* **1993**, *20*, 3.
29. Grabka, *J.; Gaz. Cukrow* **1996**, *104*, 121.
30. Kobori, S.; Japanese Patent 04,187,694, **1992**, *Chem. Abstr. 118*, 6010.
31. Kaiser, H.; Nitsch, C.; Kurzendoerfer, C-P.; Wahle, B.; Baehr, B-D.; von Delden, H. German Patent 4,416,208, **1995**, *Chem. Abstr. 124*, 59011.
32. Von Raven, A. German Patent 4,410,662, **1995**, *Chem. Abstr. 123*, 344924.
33. M. Takama, M. Adachi and M. Konagai Japanese Patent, 02259098, **1990**, *Chem. Abstr. 115*, 54430.
34. Uehara, T.; Tajima, Y. Japanese Patent 06,116,428, **1994**, *Chem. Abstr. 121*, 207219.
35. Sung, E-Y.; Kottwitz, B.; Schwadtke, K.; Sttevermann, B. German Patent 4,300,659, **1994**, *Chem. Abstr. 121*, 258,587.

36. Kottwitz, B.; Upadek, H. German Patent 4,228,044, **1994**, *Chem. Abstr. 121*, 38096.
37. Mao, M. H. K.; Bernadino, L. W.; Bissett, D. L.; Fisher, P. European Patent 193,386, **1986**, *Chem. Abstr. 106*, 6746.
38. Best, J.; Gibson, W. T.; Wiechens, J. W. European Patent 489,581, **1992**, *Chem. Abstr. 117*, 178087.
39. Urano, T; Myaji; H.; Maezato, *K.* Japanese Patent, 04,214,087, **1992**, *Chem. Abstr. 118,* 21671.
40. Ookubo, M. Japanese Patent 04,202,594, **1992**, *Chem. Abstr. 117,* 174392d.
41. Ritter, G. U.S. Patent 5,474,703, **1995**.
42. Okubu, M. Japanese Patent 04,362,078, **1992**, *Chem. Abstr.117*, 218733.
43. Kiely, D. E. U. S. Patent 5,478,374, **1995**.
44. Kiely, D. E; Chen, L.; Morton, D. W. U. S. Patent 5,434,233, **1995**.
45. Carter, R. *Industry Newsline* February 2000, www.impomag.com.

Chapter 7

Levoglucosenone: A Chiral Building Block with a New Perspective

Zbigniew J. Witczak

Department of Pharmaceutical Sciences, School of Pharmacy, Wilkes University, Wilkes-Barre, PA 18766

Levoglucosenone, a chiral bicyclic enone with excellent reactivity and functionality, is a convenient precursor for the synthesis of many attractive templates of complex, natural products and is attracting steady interest due to its rigidity and stereoselective functionalization in many synthetic organic methodologies.

Historical Background

Levoglucosenone *(1)*, an attractive chiral carbohydrate building block, is conveniently produced by the pyrolysis of cellulose composed materials. Despite the disadvantages of its low yield and the amount of solid cellulosic material necessary for pyrolysis, the efficiency and the economy of the pyrolysis process makes it an effective method. In addition, pyrolysis reduces the amount of waste cellulosic material, which is beneficial to the environment.

Although levoglucosenone has been known for more than 30 years *(2)*, it continues to have only limited applications in organic synthesis. This can be attributed to the rather conservative opinion regarding the process, purification and stability, etc.

This simple and small bicyclic enone molecule is an important and efficient chiral starting material for the synthesis of many analogs of complex natural products. Despite the efforts of various laboratories *(1,3-6)* to promote the chemistry of levoglucosenone, the interest of pharmaceutical industries in the chemical remains low. We hope continuous promotion of this remarkable molecule will make levoglucosenone a commodity product, a status that should have been granted to this molecule long, long ago. Thus, the goal of this chapter is to put levoglucosenone on the map as a valuable chiral building block.

Early Structural Studies and Physicochemical Properties

This small molecule with remarkable potential applications in synthetic organic chemistry first attracted the attention of the chemists in the early 70s. Since then, detailed structural studies have been published *(1-5)* and the revision of the previously published data for the MS and ^{1}H NMR spectra by Broido and

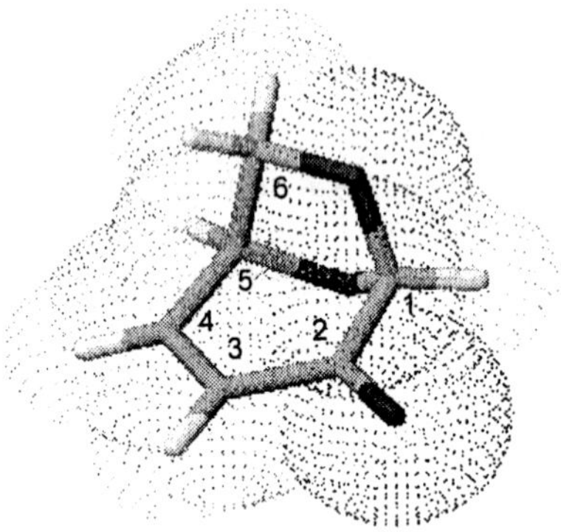

coworkers *(6-7)* clearly established its structure as 1,6-anhydro-3,4-dideoxy-α-D-*glycero*-hex-3-enopyranose-2-ulose. These studies were based on the combined GC/MS using both electron impact (EI) and chemical ionization (CI) techniques and allowed for the determination of the correct molecular ion at m/z 126.

Among the first attempts to elucidate the structure of levoglucosenone through ^{1}H and ^{13}C NMR spectral analysis was with Broido data. The ^{13}C NMR chemical shifts are shown in Scheme 1.

Interesting data on optical rotatory dispersion and circular dichroism were reported by Ohnishi *(8)* whereas Domburg *(9,10)* reported conformational and structural studies.

Halpern and Hoppech conducted a detailed NMR study of the levoglucosenone and its functionalized derivatives*(11)* and 1,4 adducts.

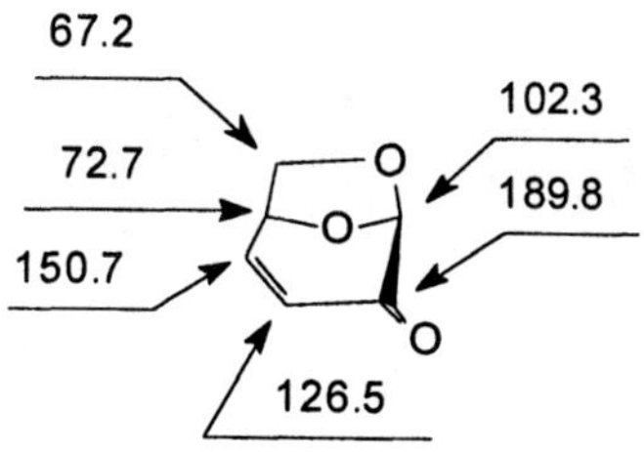

Scheme 1.

These early studies are in agreement with our structural studies in determination of torsional angels and bond distances as calculated by ACD software 3D Viewer, Version 4.5 (Advanced Chemistry Development Inc. Toronto, Canada) and depicted in Scheme 2.

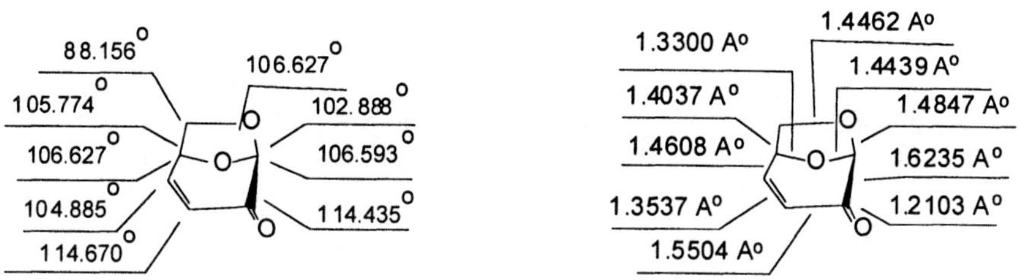

Scheme 2.

The torsional angle of the C-6 anhydro bridge is significantly lower than the corresponding C-5 and indeed correlates with the similar structural data calculated for levoglucosan derivatives *(1)*. Interestingly, the distances of the double bond of the conjugated system and double bond of the keto function at C-2 are significantly different. This is probably an important structural feature responsible for the overall reactivity of this molecule.

One of the earlier considerations in the structure elucidation and fragments analysis was levoglucosenone's photochemical degradation or photolysis. This type of study provided important data about the stability of the molecule and its decomposition products under high temperature and pressure conditions. Early reports conclusively pointed out the pattern of the fragmentation. The degradation product arose from α–cleavage of the 1,6-anhydro ring.

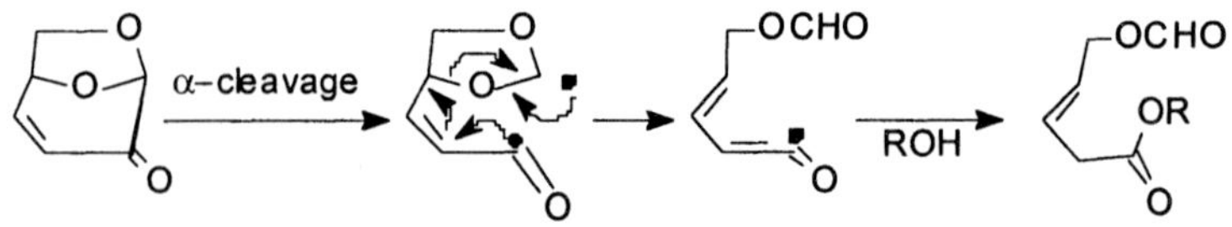

Scheme 3.

Indeed, Yamada and Matsumoto *(12,13)* reported the photochemical α-cleavage of levoglucosenone and pointed out the general pathway of the photolysis and its application to the synthesis of intermediates as convenient chiral building blocks. The intermediate diradical generated by the α-cleavage via the triplet excited state rearranges to the Z-ketene, which reacts with solvents such as water, methanol or t-butyl alcohol to form acids or esters.

Mechanism of the Formation of Levoglucosenone

A mechanism for acid catalysed thermal decomposition of cellulose *(15-26)* is a 1,2-hydride shift from the C-3 to the carbenium center at C-2 with the formation of a more stable hydroxycarbenium ion as depicted in Scheme 4. The intermediate levoglucosan is formed first *via* formation of 1,6-anhydro ring.

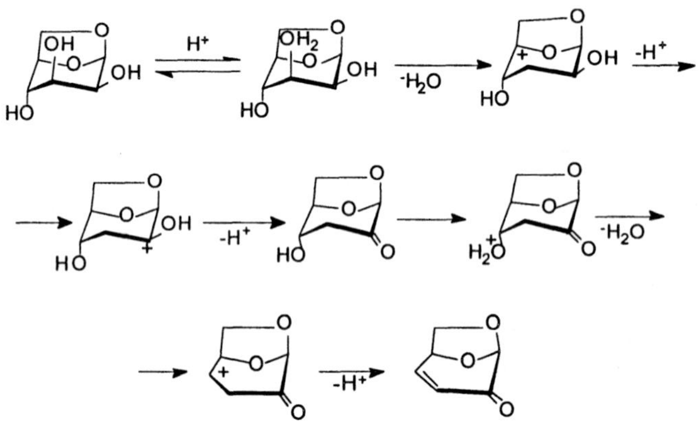

Scheme 4.

An alternative 1,2-hydride shift leading to a hydroxycarbenium ion at C-4 does not occur since the corresponding levoglucosenone isomer known as isolevoglucosenone was not found in the pyrolysate. The hypothethical mechanism of the formation of levoglucosenone via three alternative routes is depicted in Scheme 5 *(21)*.

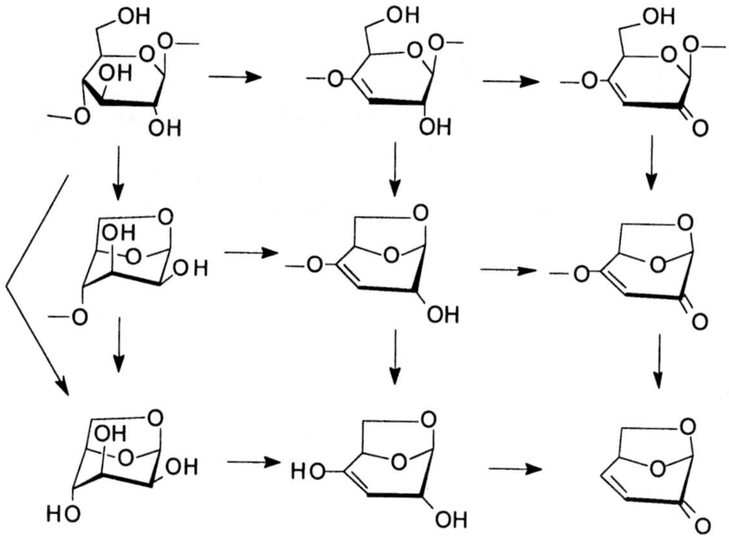

Scheme 5.

Synthesis of Levoglucosenone

Koll and coworkers *(27)* reported the first synthesis of levoglucosenone as part of a study on the utilization of 1,6-anhydrosugars in the synthesis of convenient derivatives of 1,6-anhydrosugars (Scheme 6).

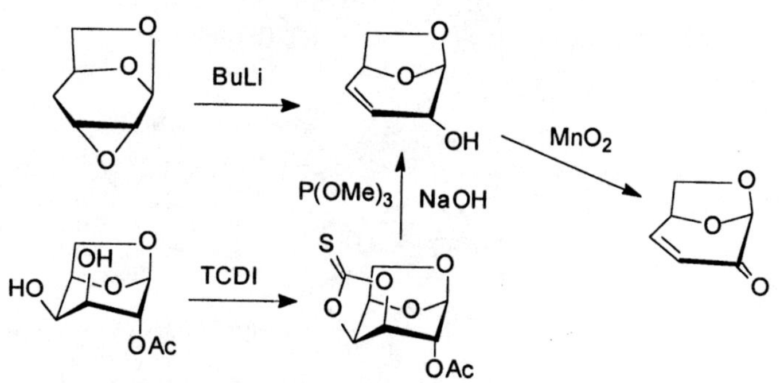

Scheme 6.

The key step proceeds *via* rearrangement of the epoxide to the intermediate allyl alcohol which on oxidation produces levoglucosenone. The most efficient approach uses a 1,6-anhydrogalactopyranose precursor through functionalization with thiocarbonyldiimidazole (TCDI) followed by desulfurization to the previously prepared allyl alcohol, which on oxidation with manganese dioxide produces levoglucosenone.

Shibagaki and coworkers *(28)* reported the second practical and efficient approach to the synthesis of levoglucosenone utilizing a galactose derivative as a starting material. This route (Scheme 7) utilized oxidative decarboxylation of 2,3-functionalized orthoesters with zirconium dioxide as a key step.

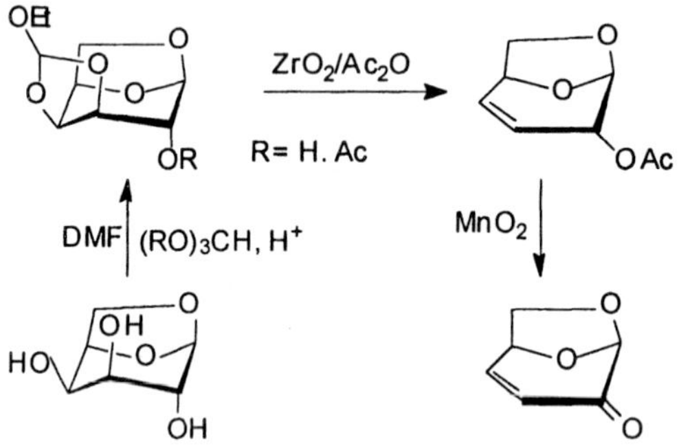

Scheme 7.

In Gallagher's herbicidin synthesis *(29)* an interesting serendipitous discovery explored a possible alternative route to levoglucosenone. A bicyclic ketone (prepared from 1,5-anhydro-D-mannitol) underwent efficient silylation and gave the silyl enol ether, which when treated with Lewis acid promoters ($TiCl_4$, $ZnBr_2$, $TMSOSO_2CF_3$ or $LiClO_4$) gave levoglucosenone. Examining the crude reaction mixtures by ^1H-NMR showed that in all cases extensive rearrangement of silyl enol ether had taken place. The researchers were able to isolate and characterize levoglucosenone as the major product. (Scheme 8)

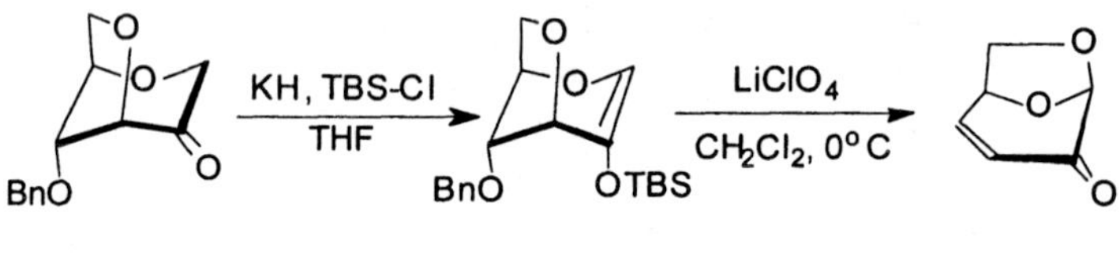

Scheme 8.

A cleaner process was observed when the silyl enol ether was treated with $LiClO_4$ *in the absence* of an additional electrophilic component and, under these conditions, levoglucosenone was isolated in 40 % yield (Scheme 8).

Our laboratory *(30)* was the first to synthesize the (+)-enantiomer of levoglucosenone and its 5-hydroxymethyl analog, starting from the known precursor, 5-hydroxymethyl-1,6-anhydro-α-*altro*-hexopyranose as depicted in Scheme 9.

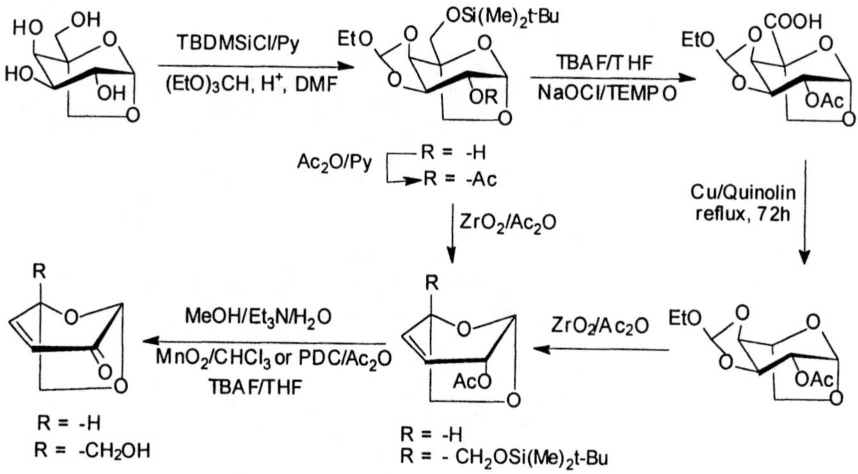

Scheme 9.

Reactivity of the Conjugated System

The high chemical reactivity of the conjugated system in levoglucosenone is an excellent reason to further develop new synthetic approaches for the synthesis of a variety of natural products targets that require stereoselective coupling with the sugar unit. As levoglucosenone is by far the most prominent carbohydrate molecule used in conjugate addition reactions, some of its tandem reactions involving the initial conjugate addition will be discussed in separate sections.

Shafizadeh's laboratory *(31-34)* reported the first successful studies of thiol addition. Our recent studies *(35-37)* and others *(38)* on the Michael addition reaction of reactive thiols to levoglucosenone and its isomer isolevoglucosenone *(39)* clearly indicate the complete stereoselectivity of the addition and its general synthetic approach to 1,4- and 1,2-thiodisaccharides *(40)*.

Reactivity of the Keto Function

The reactivity of the carbonyl group in the conjugated system of levoglucosenone is well defined and it has been shown that the carbonyl group and the double bond can be reduced selectively *(19,33)*.

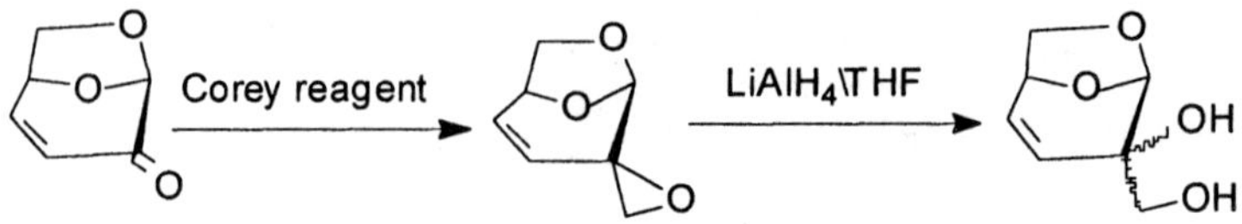

Scheme 10.

One of the important functionalizations of the keto function is the epoxidation using the Corey reagent (dimethylsulfoniumethylide in DMSO and THF) as reported by Gelas and Gelas *(41,42)* (Scheme 10).

Cycloaddition

This type of reaction has been explored in detail as the chemical character of the conjugated system and reactivity of levoglucosenone is highly suitable for functionalization through cycloaddition reactions. Horton's laboratory *(43,44)* was the first to explore the potential of levoglucosenone as a chiral dienophile in Diels-Alder cycloadditions.

Paton and coworkers *(45)* reported a highly regio- and stereospecific cycloaddition reaction of benzonitrile oxide to levoglucosenone leading to the formation of two *exo*-isomers and the *endo*- adduct in 70 % yield. A similar approach to functionalized levoglucosenone was based on the stereoselective construction of fused heterocyclic systems *(45-49)*.

Annelation

From the first attempt to reevaluate the high potential of levoglucosenone as the perfect model molecule for various types of stereoselective annulations came from the Isobe approach to reserpine *(50)*, which is a classic example of assembling the required stereochemistry as well as a specific ring sequence. This is illustrated in Scheme 11.

3-Bromo-, 3-Iodo-, and 3-Nitrolevoglucosenone Derivatives

The bromination of levoglucosenone was studied in detail *(51)*. The predominant formation of 3-bromoglucosenone was always observed. Reactivity of levoglucosenone as a dienophile in the Diels-Alder cycloaddition may be improved by introducing an electronegative group such as a halogen or nitro group. Addition of bromine to levoglucosenone and concomitant elimination of hydrogen bromide with triethylamine facilitated a one pot synthesis of 3-bromo-levoglucosenone (Scheme 12).

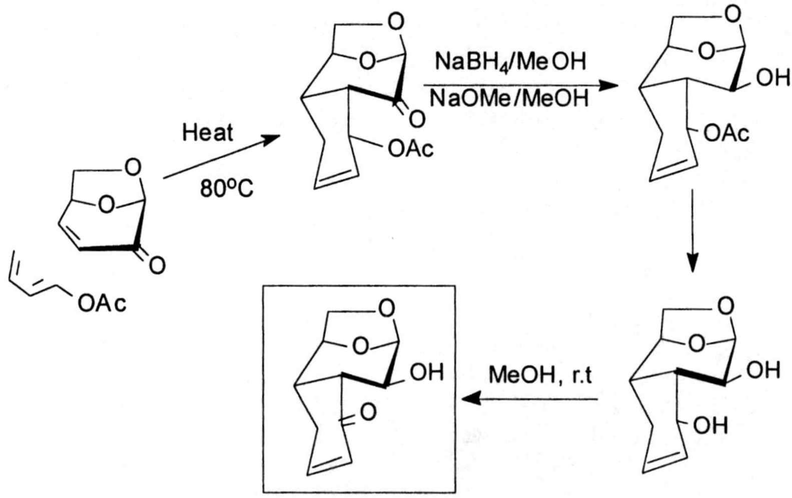

Scheme 11.

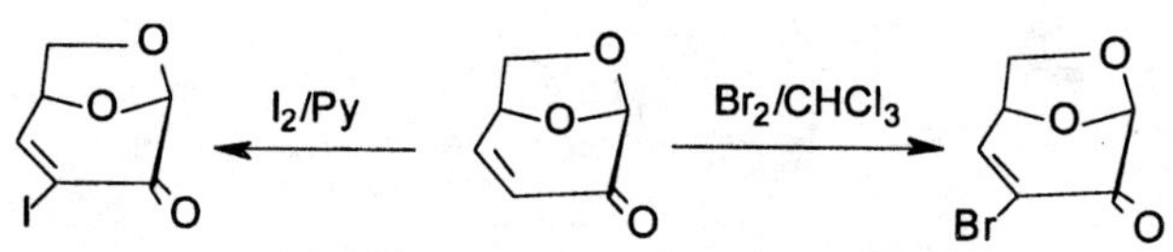

Scheme 12.

Addition of iodine to levoglucosenone has been performed conveniently by the treatment of this enone with a solution of iodine in anhydrous pyridine *(52)* resulting in the formation of 3-iodolevoglucosenone in moderate (55%) yield.

The 3-nitro analog was also synthesized by Isobe laboratory with the intention of using it as another convenient chiral dienophile in highly stereoselective cycloaddition approaches to heterocyclic systems of natural products.

These halogen derivatives have high functionality and are as equally reactive as levoglucosenone. They have been used as convenient Diels-Alder dienophiles in the Isobe synthesis of tetrodotoxin *(53)*.

Interestingly, the photobromination of 1,6-anhydro sugars (levoglucosan) developed by Ferrier's group *(54-57)* proceeds stereo- and regioselectively by substitution of H-6.

Natural Product and Rare Sugar Synthesis

Some interesting examples of levoglucosenone's application in the synthesis of natural products and rare carbohydrates have been reported *(58-81)*. Indeed, levoglucosenone has been used in the synthesis of (+)-multistriatin *(58,72-73)*, Prelog-Djerassi lactonic acid *(58,59)* and (-)-*allo*-yohimbane *(61)* The synthesis of indole alkaloid reserpine *(61)*, and serricornin *(58)*, as well as tetrodotoxin *(53,62)* were also reported from levoglucosenone or its functionalized derivatives and was reviewed earlier by us *(1)*.

A particularly interesting approach to the 1,6;3,4-dianhydro-D-talopyranose was reported by Matsumoto and coworkers *(63)*. The reaction sequence starts from the allylic alcohol (reduced levoglucosenone), which upon treatment with iodine and silver acetate in acetic acid, followed by heating with water and neutralization with methanolic ammonia, forms two compounds (Scheme 13).

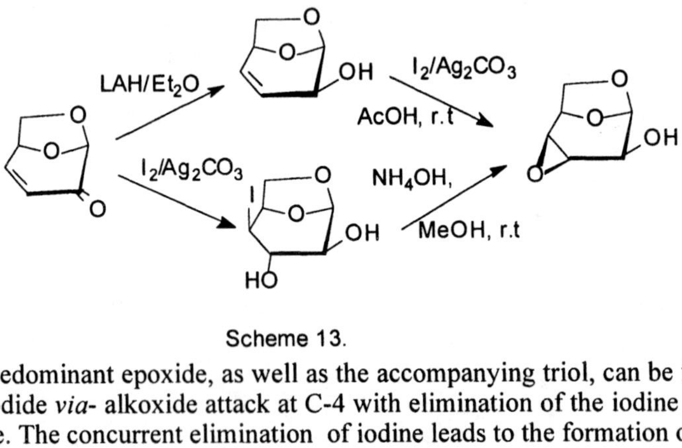

Scheme 13.

The predominant epoxide, as well as the accompanying triol, can be formed from the iodide *via*- alkoxide attack at C-4 with elimination of the iodine to give the epoxide. The concurrent elimination of iodine leads to the formation of triol.

Miscellaneous Reactions

Levoglucosenone's conjugated system undergoes base catalysed oligomerization *(64-69)* with the formation of variety of complex dimers, trimers and oligomers.

A number of other applications of levoglucosenone in the synthesis of various natural products includes (-) eldanolide *(70)*, (-) trans-cognac lactone *(71)* and (-) trans-whisky lactone *(71)*. A Baeyer-Villiger oxidation is a important key step common to all the approaches.

The improved synthetic procedure for the synthesis of δ-multistriatin *(72)* was reported from our laboratory *(73)* utilizing previously developed Paton protocol for functionalization of levoglucosenone with nitromethane *(74,75)* (Scheme 14).

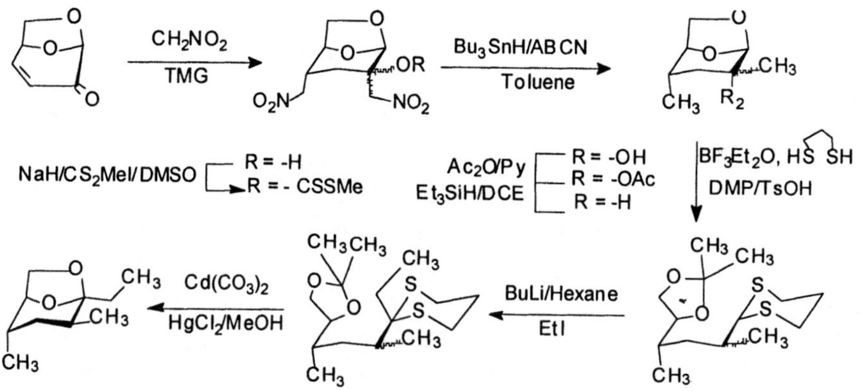

Scheme 14.

New perspectives

Although developments in the chemistry of levoglucosenone (during the last ten years) that used modern reagents as tools in organic synthesis presented in this short review will definitely change our perception of levoglucosenone's growing potential, its aggressive promotion must continue. This will encourage more extensive study in many different directions. Moreover, the excellent chiral functionality of levoglucosenone and its functionalized new synthons will create additional possibilities of interdisciplinary approaches not only in pure synthetic organic chemistry but also in polymer and combinatorial chemistry. The latter is especially appealing for functionalizing this molecule by creating a number of useful scaffolds. The most useful scaffolds would have modified functional groups such as $-NH_2$, -COOH, - SH, at C-2, C-3, C-4, and C-6. Our laboratory is developing a new family of levoglucosenone-based scaffolds with these functional groups at the above positions. (Scheme 15).

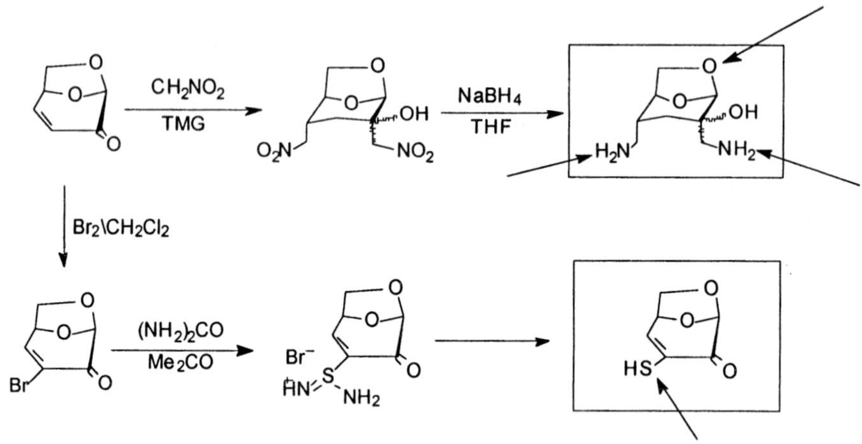

Scheme 15.

Additionally, convenient conversion of levoglucosenone into isomeric isolevoglucosenone *(82)* (Scheme 16) utilizes the more of the potential of this molecule.

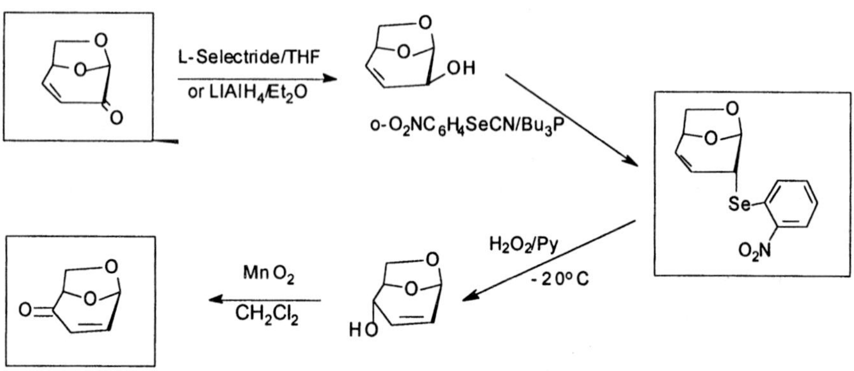

Scheme 16.

The stereoselective, one-step synthesis of (1,4) *(36,37)* and (1,2)-3-deoxy-thiodisaccharides *(39,40)* is a classical example (Scheme 17) of exploiting the excellent functionality of both levo- and isolevoglucosenone. Many other laboratories *(83-102)*, through significant contributions to the chemistry of levoglucosenone stimulated the field and increased an interdisciplinary attempt to make clear the potential of levoglucosenone and its functionalized derivatives.

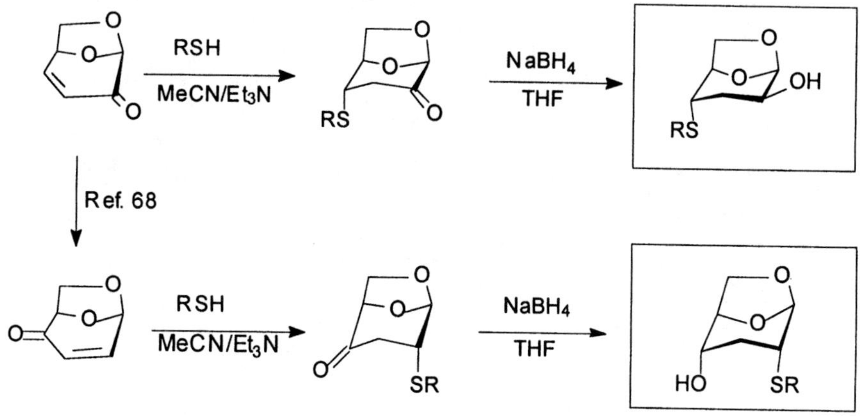

Scheme 17.

Conclusion

Through a number of new developments and synthetic methods devoted to the subject during the last ten years, one can easily conclude that this fascinating topic is growing and will continue to grow. Despite the low level of pharmaceutical industry interest, levoglucosenone chemistry will be one of the frontiers in carbohydrate chemistry, especially in the area of small molecules and complex oligosaccharides of medicinal interest. The variety of methods for the functionalization of this classical building block provides a number of attractive stereoselective approaches to various classes of optically active derivatives of particular interest including sulfur and nitrogen heterocycles as well as rare carbohydrates.

Additionally, the environmental issue of utilizing waste cellulosic material and waste biomass products should be considered as an alternative green chemistry application to the production of many value added products based on levoglucosenone. The combinatorial utilization of carbohydrate scaffolds based on levoglucosenone functionalization will also constitute attractive and relatively cheap starting materials.

This rich selection of potential approaches, combined with further developments of new procedures and modern reagents, creates an enormous opportunity for the field to be at the frontier for many years to come.

References

1. For reviews see; *Levoglucosenone and Levoglucosans Chemistry and Applications* Witczak, Z. J. Ed. ATL Press Science Publishers; Mt. Prospect, IL **1994**; Witczak, Z. J. in *Studies in Natural Products Chemistry*, Atta-Ur-Rahman, Ed. Vol. 14, Elsevier Science Publishers, Amsterdam, **1993**, pp. 267-282; Miftakhov, M. S.; Valeev, F. A.; Gaisina, I. N. *Uspekhi Khimi*, **1994**, *63*, 922; B. Becker, *J. Carb. Chem.* **2000**, *19*, 253. Trahanovsky, W.S.; Ochoda, J.M.; Wang, C.; Chang S. 217th Am. Chem Soc. Meeting, **1999**, Abstract FUEL 35.
2. Tsuchiya, Y.; Swami, K. *J. Appl. Polym. Sci.* **1970**, *14*, 2003.
3. Lam, L. K. M.; Fungi, D. P. C.; Tsuchiya, Y.; Swami, K. *J. Appl. Polym. Sci.* **1970**, *17*, 391.
4. Woodley, F. A. *J. Appl. Polym. Sci.* **1971**, *15*, 835.
5. Lipska, A. E.; McCasland, G. E. *J. Appl. Polym. Sci.* **1971**, *15*, 419.
6. Halpern, Y.; Riffer, R.; Broido, A. *J. Org. Chem.* **1973**, *38*, 204.
7. Broido, A.; Evett, M.; Hodges, C. C. *Carbohydr. Res.* **1975**, *44*, 267.
8. Ohnishi, A.; Takagi, E.; Kato, K. *Bull. Chem. Soc. Jpn.*, **1975**, *48*, 1956.
9. Domburg, G.; Berzina, I.; Kupce, E.; Kirshbaum, I. Z. *Khim. Drev.* **1980**, 99.
10. Domburg, G.; Berzina, I.; Kirshbaum, I. Z.; Gavars, M. *Khim. Drev.* **1978**, 105.
11. Halpern, Y.; Hoppech, J. P. *J. Org. Chem.* **1985**, *50*, 1556.
12. Yamada, S.; Matsumoto, M. *Chem. Lett.* **1992**, 2273.
13. Matsumoto, M. *J. Chem. Soc. Chem. Commun.* **1984**, 658.
14. Shafizadeh, F.; Fu, Y. L. *Carbohydr.Res.* **1973**, *29*, 113.
15. Shafizadeh, F.; Chin, P. P. S. *Carbohydr.Res.* **1976**, *46*, 149.
16. Shafizadeh, F.; Chin, P. P. S. *Carbohydr.Res.* **1977**, *58*, 79.
17. Koll, P.; Metzger, P. *Angew. Chem.* **1978**, *90*, 802.
18. Shafizadeh, F.; Furneaux, R. H.; Stevenson, T. T. *Carbohydr. Res.* **1979**, *71*, 169.
19. Ward D. D.; Shafizadeh F. *Carbohydr. Res.* **1981**, *93*, 284.
20. Furneaux, R. H.; Gainsford, G. T.; Shafizadeh, F.; Stevenson, T. T. *Carbohydr. Res.* **1989**, *146*, 113.
21. Shafizadeh, F. *Pure Appl. Chem.*, **1983**, *55*, 705-720; Shafizadeh, F. *J. Anal. Appl. Pyrolysis*, **1982**, *3*, 283.
22. Smith, C. Z.; Chum, H. L.; Utley, J. H. P. *J. Chem. Res. Synopses* **1987**, *3*, 88.
23. Shafizadeh, F.; Ward, D. D.; Pang, D. *Carbohydr. Res.* **1982**, *102*, 217.
24. Shafizadeh, F.; Furneaux, R. H.; Pang, D.; Stevenson,T. T. *Carbohydr. Res.* **1982**, *100*, 303.
25. Furneaux, R. H.; Mason, J. M.; Miller, I. J. *J. Chem. Soc. Perkin Trans. 1* **1984**, 1923.
26. Morin, C. *Tetrahedron Lett.* **1993**, 5095.

27. Koll, P.; Schultek, T.; Rennecke, R. W. *Chem. Ber.* **1976**, *109*, 337.
28. Shibagaki, M.; Takahashi, K.; Kuno, H.; Honda, I.; Matsushita, H. *Chem. Lett.* **1990**, 307.
29. Griffin, A.; Newcombe, N. J.; Gallagher, T.; In *Levoglucosenone and Levoglucosans Chemistry and Applications* Witczak, Z. J. Ed. ATL Press Science Publishers; Mt Prospect, IL 1994; p.23.
30. Witczak, Z. J.; Mielguj, R. *Synlett* **1996**, 108.
31. Ward, D. D.; Shafizadeh, F. *Carbohydr. Res.* **1981**, *95*, 155.
32. Shafizadeh, F.; Essig, M. G.; Ward, D. D. *Carbohydr.Res.* **1983**, *114,* 71.
33. Essig, M. G.; Shafizadeh, F. *Carbohydr. Res.* **1984**, *127*, 235.
34. Essig, M. G. *Carbohydr. Res.* **1986**, *156,* 225.
35. Witczak, Z. J. *Pure Appl. Chem.* **1995**, *66,* 2189.
36. Witczak, Z. J.; Chhabra, R.; Chen, H.; Xie, Q. *Carbohydr. Res.* **1997**, *301*, 167.
37. Witczak, Z. J.; Sun, J.; Mielguj, R. *Bioorg. Med. Chem. Lett.* **1995**, *5*, 2169.
38. Becker, T.; Thimm, B.; Thiem, J. *J. Carbohydr. Chem.* **1996**, *15*, 1179.
39. Witczak, Z. J.; Chen, H.; Kaplon, P. *Tetrahedron: Asymmetry,* **2000**, *11*, 519.
40. Witczak, Z. J.; Chhabra, R.; Boryczewski, D. *J. Carbohydr. Chem.* **2000,** *19*, 543.
41. Gelas-Mialhe, Y.; Gelas, J.; Avenel, D.; Brahmi, R.; Gillie-Pandraud, H. *Heterocycles* **1986**, *24*, 931.
42. Gelas-Mialhe, Y.; Gelas, J. *Carbohydr.Res.* **1990**, *199,* 243.
43. Bahte, P.; Horton, D. *Carbohydr. Res.* **1983**, *122,* 189.
44. Bahte. P.; Horton, D. *Carbohydr. Res.* **1985**, *139,* 191.
45. Blake, A .J.; Forsyth, A. C.; Paton, M. R. *J. Chem. Soc. Chem. Commun.* **1988**, 440.
46. Dawson, I. M.; Johnson, T.; Paton, M. R.; Rennie, R.A.C. *J. Chem. Soc., Chem. Commun.* **1988**, 1339.
47. Blake, A. J.; Dawson, I. M.; Forsyth, A. C.; Gould, R.O.; Paton, M. R.; Taylor, D. *J. Chem. Soc. Perkin Trans. 1* **1993**, 75.
48. Blake, A. J.; Cook, T. A.; Forsyth, A. C.; Gould, R. O.; Paton, R. M. *Tetrahedron* **1992**, *48*, 8053.
49. Blake, A. J.; Gould, R. O.; Paton. R. M.; Taylor, P. G. *J. Chem. Res. Synopses* **1993**, 289.
50. Isobe, M.; Fukami, N.; Nishikawa,T.; Goto, T. *Heterocycles* **1987**, *25*, 521.
51. Ward, D. D.; Shafizadeh, F. *Carbohydr. Res.* **1981**, *93*, 287.
52. Bamba, M.; Nishikawa, T.; Isobe, M. *Tetrahedron Lett.* **1996**, 37, 8199.
53. Isobe, M.; Fukuda, Y.; Nishikawa, T.; Chabert, P.; Kawai, T.; Goto, T. *Tetrahedron Lett.* **1990**, 3327.
54. Chew, S.; Ferrier, R. J. *J. Chem. Soc. Chem. Commun.* **1984**, 911.
55. Chew, S.; Ferrier, R. J.; Sinnwell, V. *Carbohydr. Res.* **1988**, *174*, 161.
56. Ferrier, R.; Furneaux, R. *Aust. J. Chem.* **1980**, *33,* 1025.
57. Ferrier, R.; Tyler, P. C. *J. Chem. Soc. Perkin Trans.1* **1980**, 2767.
58. Mori, M.; Chuman, T.; Kato, K.; Mori, K. *Tetrahedron Lett.* **1982**, 4593.

59. Isobe, M.; Nishikawa, T.; Yamamoto, N.; Tsukiyama,T.; Ino, A.; Okita, T. *J. Het. Chem.* **1992**, *29,* 619.
60. Isobe, M.; Ichikawa, Y.; Goto, T. *Tetrahedron Lett.* **1981**, 4287.
61. Isobe, M.; Fukami, N.; Goto, T. *Chem. Lett.* **1985**, 71.
62. Isobe, M.; Nishikawa, T.; Pikul, S.; Goto, T. *Tetrahedron Lett.* **1987**, 6485.
63. Matsumoto, K.; Ebata, T.; Koseki, K.; Okano, K.; Kawakami, H.; Matsushita, H. *Heterocycles* **1992**, *34*, 1935.
64. Stevenson, T. T.; Furneaux, R. H.; Pang, D.; Shafizadeh, F.; Jensen, L. Stenkamp, R. E. *Carbohydr. Res.* **1983**, *112*, 179.
65. Stevenson, T. T.; Stenkamp, R. E.; Jensen, L. H.; Shafizadeh, F.; Furneaux. R. H. *Carbohydr. Res.* **1983**, *104*, 11.
66. Stevenson, T. T.; Essig, M. G.; Shafizadeh, F.; Jensen, L. H.; Stenkamp, R. E. *Carbohydr. Res.* **1983**, *118*, 261.
67. Essig, M. G.; Shafizadeh, F.; Cochran, T. G.; Stenkamp R. E. *Carbohydr. Res.* **1984**, *129*, 55.
68. Essig, M. G.; Stevenson, T. T.; Shafizadeh, F.; Stenkamp, R. E.; Jensen, L. H. *J. Org. Chem.* **1984**, *49*, 3652.
69. Kettmann,V.; Bystricky, S.; Sticzay, T.; Koos. M. *Acta Crystallographica Sec. C. Crystal Structure Commun.* **1989**, *45,* 1580.
70. Matsumoto, K.; Ebata, T.; Koseki, K.; Okano, K.; Kawakami, H.; Matsushita, H. *Carbohydr. Res.* **1993**, *246*, 345.
71. Ebata, T.; Matsumoto, K.; Yoshikoshi, H.; Koseki, K.; Kawakami, H.; Mashushita, H. *Heterocycles,* **1990**, *31*, 423.
72. Mori, M.; Chuman, T.; Kato, K. *Carbohydr. Res.* **1984**, *129*, 73.
73. Witczak, Z. J.; Li, Y. *Tetrahedron Lett.* **1995**, 36, 2595.
74. Forsyth, A. C.; Paton, R. M.; Watt, A. *Tetrahedron Lett.* **1989**, 993.
75. Forsyth, A. C.; Gould, R. O.; Paton, R. M.; Sadler, I. H.; Watt, I. *J. Chem. Soc., Perkin Trans.1* **1993**, 2737.
76. Matsumoto K.; Ebata T.; Koseki, K.; Kawakami, H.; Matsushita H. *Heterocycles* **1991**, *32*, 2225.
77. Matsumoto, K.; Ebata, T.; Koseki, K.; Kawakami, H.; Matsushita, H. *Bull. Chem. Soc. Jpn.* **1991**, *64,* 2309.
78. Koseki, K.; Ebata, T.; Kawakami, H.; Mashushita, H.; Hoshitake N.; Itoh, K. *Heterocycles* **1990**, *31*, 1585.
79. Freskos, J. N.; Swenton J. *J. Chem. Soc. Chem. Commun.* **1984**, 658.
80. Swenton, J. S.; Freskos, J. N.; Dalidowicz, P.; Kerns. M. L. *J. Org. Chem.* **1996**, *61*, 459.
81. Taniguchi, T.; Nakamura, K.; Ogasawara, K. *Synlett* **1996**, 971.
82. Witczak, Z. J.; Kaplon, P.; Kolodziej, M. *J. Carbohydr. Chem.* **2000**, *19*, in press.
83. Tolstikov, G.A.; Valeev, F.A.; Gareev, A. A.; Khalilov, L. M.; Miftakhov, M.S. *Zh. Org. Khim.* **1991**, *27*, 565.
84. Laikhter, A. L.; Niyazymbetov, M. E.; Evans, D. H.; Samet, A.V.; Semenov, V. V. *Tetrahedron Lett.* **1993**, *34*, 4465.

85. Matsumoto, K, Ebata, T.; Koseki, K.; Okano, K.; Kawakami, H. Matsushita, H. *Carbohydr. Res.* **1993**, *246*, 345.
86. Efremov, A. A.; Slaschinin, G. A.; Korniyets, E. D.; Sokolenko,V.A.; Kuznetsov, B. N. *Sibirskii Khim. Zhur.* **1992**, *6*, 34.
87. Efremov, A. A.; Konstantinov, A. P.; Kuznetsov, B. N. *J. Anal. Chem.* **1994**, *49*, 742.
88. Blattner, R.; Page, D. M. *J. Carbohydr. Chem.* **1994**, *13*, 27.
89. Witczak, Z. J.; Chabra, R.; Chojnacki, J. *Tetrahedron Lett.* **1997**, *38*, 2215.
90. Samet, A. V.; Laikhter, A. L.; Kislyi, V. P.; Ugrak, B. I.; Semenov, V. V. *Mendeleev Commun.* **1994**, 134.
91. Niyazymbetov, M. E.; Laikhter, A. L.; Semenov, V. V.; Evans, D. H. *Tetrahedron Lett.* **1994**, *35*, 3037.
92. Matsumoto, K.; Ebata, T.; Matsushita, H. *Carbohydr. Res.* **1995**, *279*, 93.
93. Taniguchi, T.; Ohnishi, H.; Ogasawara, K. *Chem. Commun.* **1996**, 1477.
94. Samet, A. V.; Kislyi, V. P.; Chernyshova, N. B.; Reznikov, D. N.; Ugrak, B. I.; Semenov, V. V. *Rus. Chem. Bull.* **1996**, *45*, 393.
95. Samet, A. V.; Yamskov, A. N.; Ugrak, B. I.; Vorontsova, L. G.; Kurella, M. G.; Semenov, V. V. *Rus. Chem. Bull.* **1996**, *45*, 393.
96. Valeev, F. A.; Gaisina, I. N.; Miftakhov, M. S. *Rus. Chem. Bull.* **1996**, 45, 2453.
97. Valeev, F. A.; Gaisina, I. N.; Sagitdinova, K. F.; Shitikova, O. V.; Miftakhov, M. S. *Zh. Org. Khim.* **1996**, *32*, 1365.
98. Samet, A. V.; Niyazymbetov, M. E.; Semenov, V. V.; Laikhter, A. L.; Evans, D. H. *J. Org. Chem.* **1996**, 61, 8786.
99. Miftakhov, M. S.; Gaisina, I. N.; Valeev, F. A. *Rus. Chem. Bull.* **1996**, *45*, 1942.
100. Nishikawa, T.; Araki, H.; Isobe, M. *Biosci. Biotechnol. Biochem.* **1998**, *62*, 190.
101. Takeuchi, M.; Taniguchi, T.; Ogasawara, K. *Synthesis* **1999**, 341.
102. Gomez, M.; Quincoces, J.; Peske, K.: Michalik, M. *J. Carbohydr. Chem.* **1999**, *18*, 851.
103. Trahanovsky, W. S.; Revell, K. D.; Arvidson, K. B.; Wang, C.; Wang Y. 219th Am. Chem Soc. Meeting, **2000**, Abstract CARB 85.

Chapter 8

New Oligomers and Polymers Bearing Furan Moieties

Claire Coutterez, Cécile Goussé, Rana Gheneim, Sandrine Waig Fang, and Alessandro Gandini*

Ecole Francaise de Papeterie et des Industries Graphiques (INPG), BP65, 38402 Saint Martin d'Hères, France

This paper describes novel approaches to the exploitation of both furan monomers and a specific facet of furan reactivity in order to synthesize either conjugated oligomers incorporating the heterocycle in their backbone, or polymeric structures which can be crosslinked and returned to linear structures through the reversible chemistry of the Diels-Alder reaction. The first family of compounds showed interesting features in terms of conductivity, luminescence, mesogenic character and photoactivity. The second class of materials owes its interest to the possibility of recycling otherwise intractable polymers, e.g. tires, thanks to a simple thermal process.

Furan derivatives are among the best examples of a source of chemicals and materials based on renewable resources because furfural is readily produced in good yields from the hemicelluloses contained in a wide variety of agricultural and forestry by-products, through a simple and economically viable process. The use of furan monomers to synthesize polymers and copolymers and of some specific features of furan chemistry to develop original macromolecular architectures have been major research topics for many years in our laboratory which is also involved in studies dealing with other renewable resources (*1*). The synthesis of these monomers and polymers as well as their characterization, properties and possible utilizations has been extensively described in a recent monograph (*2*). Some peculiarities of the chemical behavior of the furan heterocycle which can be exploited in macromolecular engineering are also included in that review. This ongoing broad investigation strategy has recently provided some novel results which are expounded here. For the sake of comparison, some structures bearing thiophene moieties are also discussed.

Conjugated Oligomers

Within the realm of conjugated macromolecules, the working hypothesis based on "the oligomer approach" has recently acquired a high status (*3*) because it provides a rigorous way of examining in depth the precise structures of individual compounds and to establish therefrom a clear-cut correlation with the various chemical and physical properties and hence the potential applications of the ensuing materials. The oligomers themselves can be quite useful in certain areas, but moreover the thorough knowledge of their features provides a more reliable insight into the potential of the corresponding high polymers.

Several years ago, we developed an original synthetic route which allowed us to prepare poly(2,5-furylene vinylene) by the polycondensation between the methyl group and the aldehyde function of 5-methylfurfural in a strongly nucleophilic medium (*4,5*):

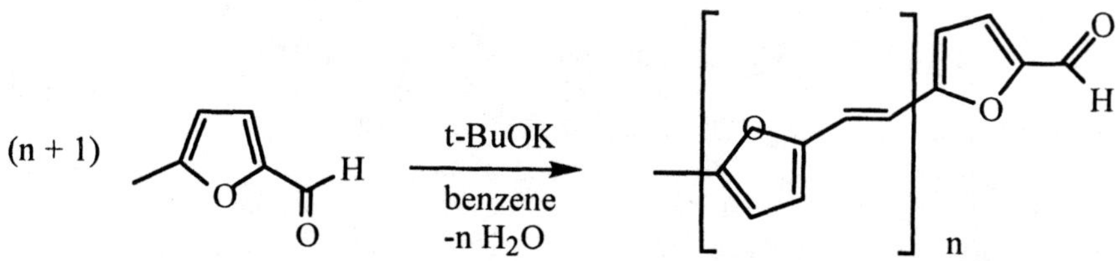

Ia-e (n=1, 2, 3, 4-8 and ~50)

This reaction is a peculiar extension of the classical aldol-crotonic condensation applied here to a methyl group which owes its capacity to lose a proton to the dienic character of the furan ring. The highest average DPn of the products obtained was about 50 (**Ie**) and their spectroscopic characterization indicated that the expected structure shown above was certainly the dominant feature, although other moieties, arising from side reactions were also clearly present, e.g., OH and COOH groups in minor proportions. However, it was the first time that this polymer had been prepared in regular linear form and its basic structure assessed spectroscopically. This furanic homologue of poly(1,4-phenylene vinylene) displayed interesting features, including a reasonable electronic conductivity and a good stability to aging, since it remained soluble in common solvents for years. 5-Methyl-2-thiophene carboxyaldehyde behaved likewise, albeit with lower rates and the polymers had lower DPs. The aromatic homologues of these heterocyclic structures did not give any condensation in the conditions adopted here. This progressive decrease in reactivity follows well the corresponding decrease in dienic character (furan > thiophene > benzene). The difficulty associated with a more detailed study of the properties of these conjugated polymers arose from the uncertainty about the precise role of spurious moieties, which were obviously impossible to "extract" from the regular polymer sequences. In order to overcome this drawback, a study of individual oligomers possessing the same repeat units became necessary.

Synthesis and Characterization

The procedures adopted to synthesize these oligomers called upon two distinct approaches. The first involved the preparation of a mixture of oligomers following our previously optimized polymerization process, but stopped in this case at a much earlier stage of polycondensation. The bulk product was submitted to a fractionation sequence based on the high-vacuum distillation of the dimer **Ia** and the subsequent separation of the trimer **Ib**, tetramer **Ic** and oligomers **Id** by flash chromatography with different solvent mixtures.

The second consisted in reacting an excess of furfural (or thiophenecarboxyaldehyde) with 5-methylfurfural (or its thiophene homologue) to yield the corresponding dimers **IIa-d**. These products were then used (in excess) to react with one further molecule of either of the two 5-methylaldehydes to obtain the trimers **IIe-h**. Finally, tetramer **IIi** was prepared from an excess of trimer **IIf** and 5-methylfurfural following the same protocol.

The purification of each compound of both series and their spectroscopic characterization, coupled with elemental analysis and DSC, showed single well defined products in all cases (*6*).

As indicated in the structures below, all these oligomers had *trans* conformations with respect to the alkenyl bridges, indicating that this mode of enchainment was always the kinetic product resulting from each condensation. The uv spectra of series **I** displayed a bathochromic shift as a function of the extent of conjugation, as shown in Table I. From these results and our previous work on higher polymers (*5*), it seems clear that beyond a DP of about 6 the planarity of these molecules is progressively lost and this leads to a limiting value of the λ_{max} at around 550 nm. Table I shows moreover that protonation of these oligomers with CF_3SO_3H produced a strong bathochromic shift in all instances, indicating that the creation of holes in the corresponding structures increases the delocalization of the electrons and thus decreases their corresponding excitation energy.

For the oligomers of series **II**, the trend was the same, as shown in Table II, which also illustrates the slight bathochromic shift associated with the replacement of a furan heterocycle with the thiophene counterpart.

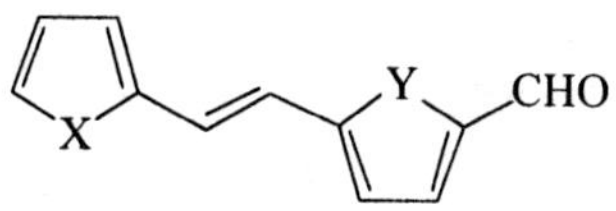

IIa: X,Y = O
IIb: X = S, Y = O
IIc: X = O, Y = S
IId: X, Y = S

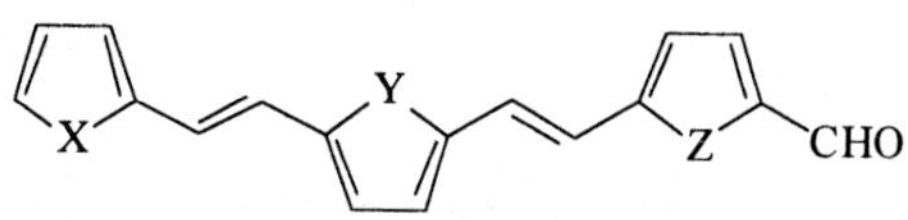

IIe: X, Y, Z = O
IIf: X = S, Y,Z = O
IIg: X, Y = O, Z = S
IIh: X, Z = O, Y = S

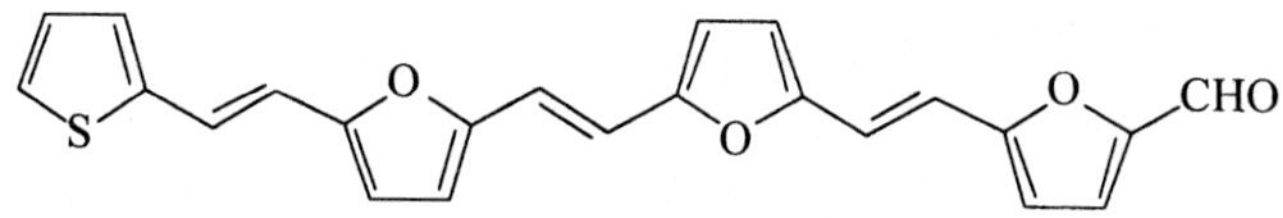

IIi

Table I. Maximum absorption wavelength and molar extinction coefficient in the UV-visible spectra (CH_2Cl_2) of series I oligomers before and after protonation

Compound	*λ_{max} (nm)*	*10^{-4} ε ($l\ mol^{-1}\ cm^{-1}$)*	*λ_{max} (nm) after protonation*
Dimer **Ia**	387	3.9	527
Trimer **Ib**	445	3.4	615
Tetramer **Ic**	475	4.1	755
Oligomers **Id** (n = 4-8)	510	nd	> 900

Table II. Maximum absorption wavelength of series II oligomers (CH_2Cl_2)

Compound	λ_{max} *(nm)*
Dimer **IIa**	370
Dimer **IIb**	375
Dimer **IIc**	380
Dimer **IId**	380
Trimer **IIe**	430
Trimer **IIf**	435
Trimer **IIg**	440
Trimer **IIh**	435
Tetramer **IIi**	470, 495 (sh)

Properties

The obvious implication of the presence of planar conjugated sequences is that it might be possible to obtain electronic conductivity with these materials. Indeed, a pellet of mixed oligomers **I** (n = 4-8) exhibited, after doping with iodine, a conductivity of 0.4 S cm^{-1} at room temperature, which indicated a typical semi-conducting behavior. These good conducting properties, related to *well-defined soluble low-DP* structures, compare very favorably with those of many *ill-defined insoluble* conjugated *polymers* and open the way to their possible use as processable conducting materials.

The photoluminescence of both series of oligomers was also studied. Given the range of absorption maxima reported above, two excitation sources were chosen, namely the 366 nm line of a medium-pressure mercury lamp and the 488 nm emission of an argon-ion laser. When excited with the 366 nm line, all these compounds displayed some emission, even if the tetramer and the higher oligomers absorbed poorly at this wavelength. With the 488 nm excitation, the absence of luminescence from the dimer was simply due to the fact that this compound did not absorb above about 420 nm.. Table III gives the position of the maximum intensity in these emission spectra as recorded with both excitation wavelengths and shows that each compound that absorbed at both excitation wavelengths, gave two identical emission spectra. This indicates that the luminescence was produced from the same emitting state. Additionally, whatever the excitation wavelength, the emission λ_{max} increased with increasing DP of the excited oligomer. Indeed, the electronic emission spectra displayed a bathochromic shift of λ_{max} as a function of the degree of conjugation, as already observed for the corresponding absorption spectra.

Table III. Emission wavelength maxima of the oligomers (CH_2Cl_2)

Compound	*λ_{max} of emission (nm) (for λ_{exc} = 366 nm)*	*λ_{max} of emission (nm) (for λ_{exc} = 488 nm)*
Dimer **Ia**	493	-
Trimer **Ib**	590	590
Tetramer **Ic**	635	635
Oligomers **Id** (n = 4-8)	650	650
Dimer **IIa**	470	-
Dimer **IIb**	490	-
Dimer **IIc**	485	-
Dimer **IId**	490	-
Trimer **IIe**	565	565
Trimer **IIf**	565	565
Trimer **IIg**	595	595
Trimer **IIh**	560	560
Tetramer **IIi**	650	650

The dimers emitted in the blue or blue-green region, whereas higher oligomers glowed in the red. The emission colours of the trimer and tetramer were intermediary as indicated by the figures in Table III. A decrease in the extent of this bathochromic shift was again observed as a function of a further DP increase, which was in tune with the similar trend discussed above for the absorption spectra. Switching heterocycles also produced the same trend as observed in the absorption spectra, with a small bathochromic effect induced by the thiophene ring. It follows that the combined use of two parameters, namely the DP of the oligomers and the sequence of heterocycles along their chain, gives rise to the possibility of generating a wide range of emission wavelengths (and therefore of colours) which could find promising applications, particularly if these trends were also displayed by electroluminescence (which is under study).

This luminescence behavior was unambiguously assigned to fluorescence because of the good continuity, in all instances, between the highest absorption wavelength and the lowest emission wavelength. This indicated the existence of a 0-0 level frontier. Moreover, the presence of atmospheric oxygen in the sample solutions (permanent contact with air during experimentation) inevitably quenched any possible phosphorescence. Indeed, oxygen is a well known powerful trap of triplet excited states. Experiments carried out in an argon atmosphere did not show any appreciable difference in the emission spectra, which suggests that the contribution of phosphorescence was negligible. However, the photomultiplier used for the detection of the emitted light was not sensitive beyond 900 nm and this leaves open the possibility of phosphorescence appearing in the infrared, if the triplet states of these molecules was associated with a particularly low energy level.

The photochemical behavior of these oligomers was first examined with the dimers which were irradiated with a medium-pressure mercury lamp through a Pyrex filter to remove any wavelength below about 300 nm. In dilute methylene chloride solutions, all the dimers (**Ia** and **IIa-IId**) isomerized to the *cis* conformation until the *cis/trans* thermodynamic equilibrium mixture was reached. No other photoinduced reaction was detected in these conditions. Irradiation of concentrated solutions, as well as crystalline solid samples, gave rise to molecular dimerization through the alkenyl bridge with all the dimers, through the classical [π2+π2] cycloaddition process, e.g. for dimer **Ia**:

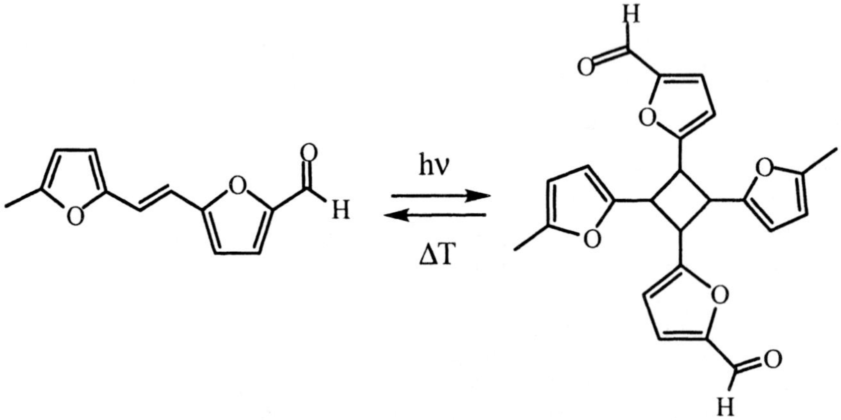

No accompanying free radical photoreactions were detected, since the course of these photolyses was the same in the presence and the absence of radical traps. The cyclobutane photodimers reverted to their precursors upon heating at 150-180°C. This clean feature was applied to the synthesis of photocrosslinkable polymer electrolytes by the following procedure. Commercial linear oligo(ethylene oxide) and star shaped oligo(propylene oxide) bearing two and three terminal primary amino groups were end-capped with dimer **Ia** with formation of the corresponding Schiff bases. Mixtures of these products were irradiated with $\lambda > 300$nm and the various photocouplings between two furanic chromophore end-groups yielded a polyether-based network which exhibited good ionic conduction after being charged with

$LiClO_4$. The interest of this material stems of course from the possibility of processing viscous liquids (the Schiff bases) into thin films containing the lithium salt before inducing the rapid photochemical cross-linking (*7*).

The trimers and tetramers do not absorb at the lower wavelengths associated with an energy range sufficient to promote cyclodimerization between an excited and a ground-state molecule. Their excitation, therefore, does not lead to any appreciable photochemistry, i.e., only photophysical processes take place to bring the excited molecules back to the ground state (emissions and non-radiative crossings).

These conjugated oligomers are likely to form organized molecular assemblies if coupled with suitable complementary structures. The dimers **Ia** and **IIa-IId** were made to react with various aromatic mono- and diamines in order to build mesogenic molecular structures capable of giving rise to thermotropic liquid crystals (*8*). When the dimer moiety was entirely furanic, no mesophase appeared, whereas the thiophenic and mixed moieties produced clear-cut nematic phases.

Exploitation of the Diels-Alder Reaction

Furan and many of its derivatives behave as dienes in the context of the Diels-Alder cycloaddition and react therefore readily with dienophiles like maleic anhydride, maleimides and propyolic esters. Perhaps the most classical example of this reaction involving the furan heterocycle is:

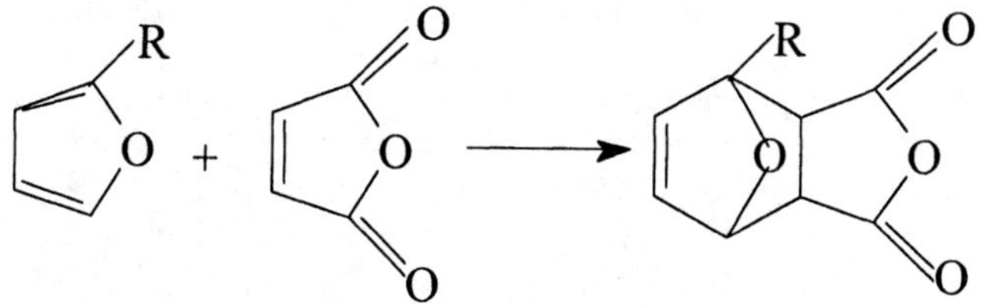

Bis-dienophiles can react with two furan moieties, e.g.:

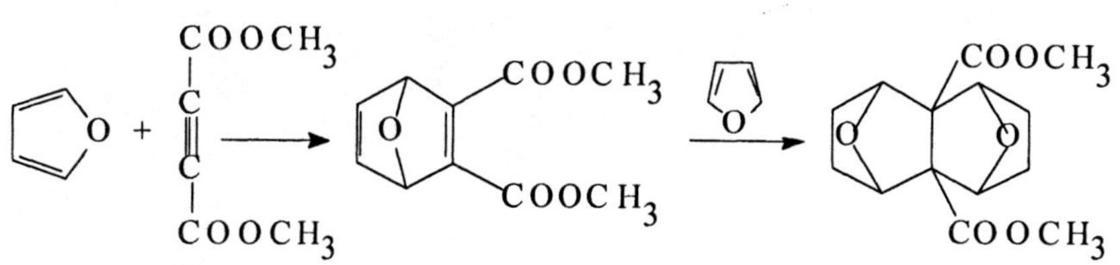

One of the peculiarities of the Diels-Alder reaction is its facile reversible character which is enhanced by a temperature increase. The application of these features to polymers bearing furan moieties has been one of the more recent research topics in our laboratory. Two different aspects related to this topic are being investigated, namely the transformation of side or terminal furan rings into more polar moieties after coupling with maleic anhydride and the reversible cross-linking of copolymers possessing the heterocycles as side groups.

An example of the first application of the Diels-Alder reaction relates to the modification of polymers with a furan end-group. These functionalized macromolecules are prepared by the cationic polymerization of a monomer in the presence of a 2-alkylsubstituted furan, by exploiting the strong reactivity of its C5 position towards electrophilic substitution. Thus, the carbenium ion chain carrier will readily couple with this position to give a polymer molecule ending with a 2,5-disubstituted furan ring. These syntheses, as well as those of block copolymers based on the same principle, have been described from 1984 (*2, 9,10*). The Diels-Alder reaction of maleic anhydride with this furan terminal group and the subsequent hydrolysis of the adduct gives rise to two COOH end-groups. The idea is to prepare non-polar oligomers or polymers like poly(isobutene) or poly(styrene) bearing a very polar extremity, which can play an interesting "surfactant" role, e.g. as compatibilizers in polymer blends. Thus for example, addition of small amounts of these materials to a mixture of a polar and a non-polar polymer would reduce their interfacial energy and therefore provide a blend with smaller microdomains. One of the tests for this strategy was the synthesis of $(COOH)_2$-terminated poly(isobutene) and the study of its mixtures with poly(acrylic acid) and poly(vinyl alcohol). In the second approach, a series of copolymers between conventional monomers (styrene, isobutene, acrylates and methacrylates) and furanic comonomers (vinyl furans, a styrylfuran and furfuryl acrylate and methacrylate) were prepared, each with varying proportions of the pending furan heterocycle, going from 3 to 50%. An example of this type of material was that resulting from the radical copolymerization of styrene with a furanic monomer synthesized in the laboratory from 4-chloromethylstyrene and 2-furfuryl alcohol. The ensuing random copolymers had the following general structures, but bore variable proportions of the comonomers:

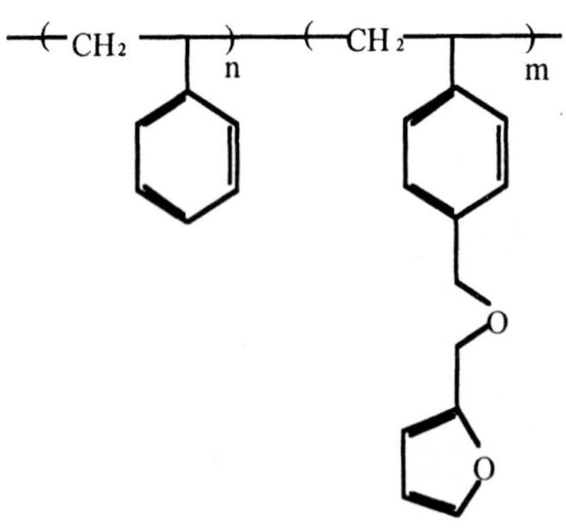

First, these copolymers were treated with a mono-maleimide in order to study the corresponding Diels-Alder reaction. The products were fully characterized because they were still soluble. Then, the same copolymers were made to react with bis-maleimides and the statistical interchain coupling through double Diels-Alder reactions produced cross-linked materials, even with as little as 3% of furan rings attached to the copolymers. Thus, the possibility of using the Diels-Alder reaction to cross-link polymers bearing small fractions of furan rings was verified and shown to be straightforward. This was found to be true also of polymers bearing the heterocycle *within* the chain (11).

These gels were then submitted to a retro-Diels-Alder reaction by suspending them in a high-boiling point solvent in the presence of an excess of 2-methylfuran and heating at 130°C for 24 h. At this temperature, the equilibrium coupling-decoupling was heavily shifted to the right hand side and the free maleimide moieties reacted with the excess of 2-methylfuran, as clearly suggested by the progressive dissolution of the polymer. After cooling, the material was precipitated and its characterization proved that the original copolymer had been restored, albeit with a few Diels-Alder adducts still pending. In this way, we showed that it is possible to decrosslink a network built through double Diels-Alder cycloadditions. In these studies, we used aromatic maleimides (*12*) which are particularly rigid structures.

In order to study elastomeric networks, simulating the type of polymers used for tires, we switched to polymers with low glass transition temperatures and oligoether bis-maleimides. A typical random copolymer structure, built from the radical copolymerization of n-hexyl acrylate and 2-furfuryl methacrylate, is shown below. These reactions were conducted in toluene at 80°C with AIBN as initiator. After 8 h, the copolymers were recovered by precipitation in 70 to 80% yields. The compositions varied from 2 to 30% of the furanic monomer (monomer feed and copolymer composition were always very similar, suggesting that r_1 and r_2 must have both been close to unity). The corresponding Tgs went from −70 to 30°C for molecular weights of about 20,000. Both homopolymers were also prepared as reference materials.

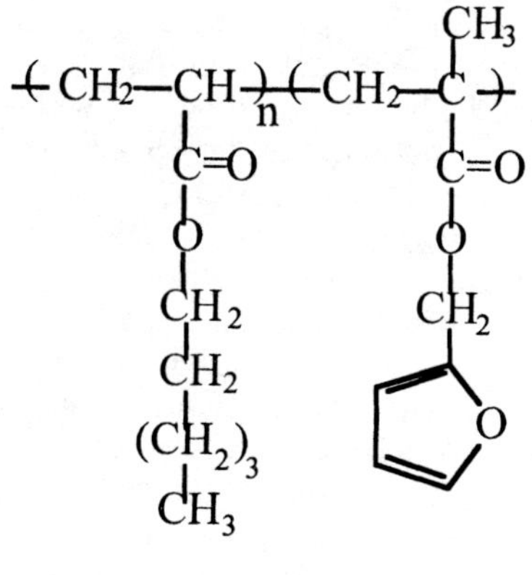

The mono-maleimide was prepared by treating an oligomeric ethylene oxide (Mn = 600) bearing one NH_2 end-group with maleic anhydride and the corresponding bis-maleimide through the same procedure, but using an oligoether (Mn = 600) with two terminal NH_2 functions. The structure of the latter is shown here:

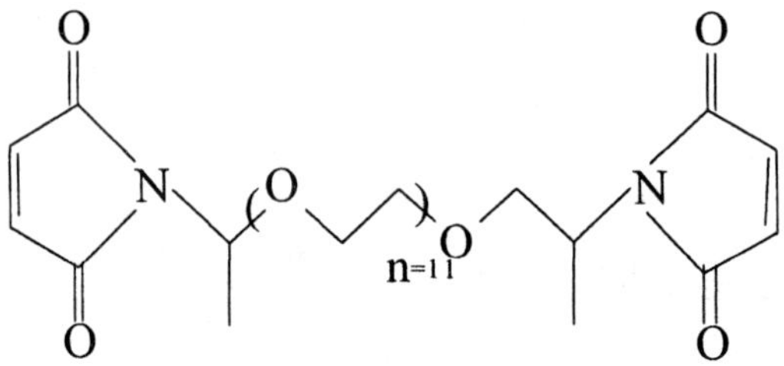

Both these products had a very low Tg, as expected from their very flexible structures.

Again, the first Diels-Alder reaction studied involved an acrylic copolymer rich in pendant furan rings and the oligoether mono-maleimide. The coupling occurred readily and in high yields. With a copolymer possessing only a few percent of heterocycles, an excess of mono-maleimide was required to induce a good yield of Diels-Alder coupling. The characterization of the various adducts arising from these cycloadditions were carried out qualitatively by FTIR and quantitatively by ^{1}H-NMR spectroscopy. Elemental analysis (particularly based on nitrogen content) and DSC completed the examination of these products.

The same procedures were applied to the combinations of the copolymers with the oligoether bis-maleimide. The formation of a gel indicated the occurrence of intermolecular couplings through the Diels-Alder reaction. The FTIR spectra of these networks and their elemental analyses clearly showed that these cycloadditions had indeed taken place. Moreover, blank experiments in the same conditions but in the absence of the bis-maleimide, did not produce any detectable change in the structure of the acrylic-furanic copolymers.

The fact of having used a bis-maleimide which was both "longer" and more flexible than the aromatic counterparts employed in our first study (*11,12*) introduced two major changes in these systems, namely a higher probability of intermolecular cycloadditions (higher mobility and wider sweep of the second maleimide group) and networks with much lower Tg, not only because the starting copolymer had chains which displayed a high flexibility, but also because the bridging structures had an equal, if not higher, softening character. Work is in progress to decrosslink these new networks by the retro-Diels-Alder reaction carried out in the presence of an excess of a suitable furan derivative, capable of trapping the maleimide functions regenerated at high temperature from the adducts. Other systems are also being inspected, with the aim of approaching the structures used in tires. It is in fact hoped that recyclable elastomers could arise from this investigation.

Conclusions

It is hoped that this brief survey of two distinct applications associated with furan monomers and furan chemistry will convey an idea of the broad field of interest associated with our strategy. The variety of monomers and architectures is very wide and consequently the potential applications of the ensuing materials range from conventional realms like polyurethane foams, polyesters, etc., to high technology areas like electronics, non-linear optics, high strength fibers, etc.

References

1. Gandini, A.; Belgacem, N. M. *Polym. Internat.* **1998**, *47*, 267.
2. Gandini, A.; Belgacem, N. M. *Progr. Polym. Sci.* **1997**, *22*, 1203.
3. *Electronic Materials: The Oligomeric Approach;* Müllen, K.; Wegner, G., Eds.; Wiley-VCH: Weinheim, FRG, 1998.
4. Gandini, A.; Méalares, C. *Trends Polym. Sci.* **1994**, *2*, 127.
5. Méalares, C.; Hui, Z.; Gandini, A. *Polymer* **1996**, *37*, 2273.
6. Coutterez, C.; Gandini A. *Polymer* **1998**, *39*, 7009.
7. Albertin, L.; Stagnaro, P.; Coutterez, C.; Le Nest, J. F.; Gandini, A. *Polymer* **1998**, *39*, 6178.
8. Coutterez, C.; Gandini, A.; Valenti, B.; Costa, G. *J. Chem. Res.* **1998**, 679.
9. Razzouk, H.; Bouridah, K.; Gandini, A; Cheradame, H. in *Cationic Polymerization and Related Processes;* Goethals, E.J., ED.; Academic Press; London, 1984; p. 355.
10. Gandini, A.; Salon, M. C.; Choura, M.; El Gharbi, R.; Amri, H.; Hui, Z. *Makromol. Chem, Macromol. Symp.* **1992**, *60*, 165.
11. Laita, H.; Boufi, S.; Gandini, A. *European Polym. J.* **1997**, *33*, 1203.
12. Goussé, C.; Hodge, P.; Gandini, A. *Macromolecules* **1998**, *31*, 314.

Chapter 9

Furfural and Levoglucosan Production from Deciduous Wood and Agricultural Wastes

Janis Gravitis[1], Nikolay Vedernikov[2], Janis Zandersons[2], and Arnis Kokorevics[2]

[1]Institute of Advanced Studies, The United Nations University, 53–67, Jingumae 5-chome Shibuya-ku, Tokyo 150–8304, Japan (email: gravitis@ias.unu.edu)
[2]Latvian State Institute of Wood Chemistry, 27 Dzerbenes Street, Riga LV1006, Latvia (email: ved@edi.lv

A new approach to increase furfural production and minimize cellulose destruction has been proposed as an alternative to the conventional dilute acid process. The new method uses small amounts of strong catalysts to provide differential catalysis of hydrolysis and dehydration reactions. The change in the process has solved two problems simultaneously: the furfural yield has increased from 55% to 75% of the theoretical and the degree of cellulose degradation has diminished fivefold. On the basis of fundamental studies, a new technology including two-step selective catalysis of wood and other pentosan-containing raw materials has been devised. This paper considers alternative routes of chemical processing of the cellulose residues into valuable monomers such as levoglucosan and ethanol as well as integration with other biomass treatment methods such as steam explosion.

Oil and Biomass Refineries

The international climate conferences in Kyoto (1997), Buenos Aires (1998) and Bonn (1999) can be regarded as tests for the human capacity to cooperate and creatively manage two dominating carbon-rich solar energy conversion products: fossil organic materials and biomass. The former is found in rich deposits and is physically rather homogeneous (oil, coal and peat), whereas the latter is widely dispersed and highly diversified (microorganisms, plants and animals). Those aspects give oil refineries the character of compact chemical plants, whereas biomass refineries (biorefineries) are just as diverse as their feedstocks (mills for grain- and oilseeds, the food industry, fermentation plants, pulp and paper mills, etc.) This situation can inspire two questions *(1)*. The first question is how can the fossil carbon sources be utilized without releasing greenhouse gases such as methane and carbon dioxide to the atmosphere? In contrast to products from non-renewable resources, wood materials do not influence the atmospheric CO_2 balance. According to *(2)*, this reduction of the atmospheric CO_2 amount to at least 187 kg for every 100 kg of dry wood. The second question is, when the oil production finally drops, can clusters of processing units, designed for the upgrading of specific bioresources, turn out a similar multitude of products as oil refineries do? To study this aspect, we shall focus on the existing and upcoming plant biomass refineries producing furfural, levoglucosan, bioethanol, etc. and their integrated production.

Industrial Uses of Furfural

In 1990, furfural plants produced not less than 240,000 tons of furfural worldwide *(3)*. The real number may exceed 300,000 tons because no information about the capacities of some furfural producing countries (under the communist regime) was available. The market price in 1990 was $0.79/lb and $0.80/lb for furfural and furfuryl alcohol, respectively.

The main furfural applications can be divided into the following groups:

- chemical intermediate (mainly as furfuryl alcohol and tetrahydrofuran)
- solvents
- others

According to *(3)*, the U.S. consumption in 1990 was 44 million lbs of furfuryl alcohol and 38 million lbs of tetrahydrofuran, while the solvent use was 8 million lbs and all others 4 million lbs.

Furfuryl alcohol is prepared by hydrogenation of furfural and is a monomer for furan resins. Industrial furan resins are corrosion-resistant, non-burning and are characterized by low smoke emission and excellent char formation. Furan polymers can also include formaldehyde, urea, phenols, etc., in their structure. The major market application of furan resins is as foundry binders for production of cores and molds used for casting metals, corrosion-resistant fiberglass-reinforced plastics, silica-filled plastics and corrosion-resistant mortars and grouts. A subsequent decrease in the price of furfuryl alcohol increases the competitiveness with phenols from oil. Moreover, the properties of furan resins are much better than those of phenol-based resins. Generally, furan resins can have a wide range of corrosion and heat properties, in comparison with the competing polymers such as polyesters, vinyl esters and epoxies. Furfuryl alcohol is an excellent solvent, a plasticizer for phenolic resins and a chemical intermediate for different syntheses.

Tetrahydrofuran (THF) is produced by decarbonylation of furfural to furan followed by catalytic hydrogenation. THF is applied as a solvent for resins and plastics, film castings and adhesives. THF also acts as a solvent in different fine organic syntheses on a commercial scale and as a chemical intermediate. Environmental demands increase the THF application cost because THF users must install solvent recovery systems. A high price and environmental considerations will limit the future growth of THF market.

Furfural solvent is used as a selective extractant for separating saturated and unsaturated hydrocarbons during the refining of lubricating oils. Furfural selectively extracts aromatic and olefinic compounds and leaves the paraffinic and naphthenic components behind. Furfural is used to remove impurities in the industrial butadiene production process.

It is used also as a secondary decolorizing agent for crude wood rosin and a selective solvent for extraction and crystallization of anthracene from crude anthracene oils. Furfural is used as an intermediate in pharmaceutical, herbicide, fragrance and flavoring applications.

Conventional Industrial Furfural Production: Unsolved Problems

In the conventional dilute sulfuric acid-catalyzed percolation process, the plant raw material containing pentosans is mixed with dilute sulfuric acid and, in the first step, pentosans are hydrolyzed to pentoses $(C_6H_{10}O_5)_n$, which are then cyclodehydrated to furfural in the second step. The furfural formed is recovered by steam distillation and fractionation. The chemical reaction is shown in Figure 1. The mechanism of the reaction *(4)* is shown in Figure 2.

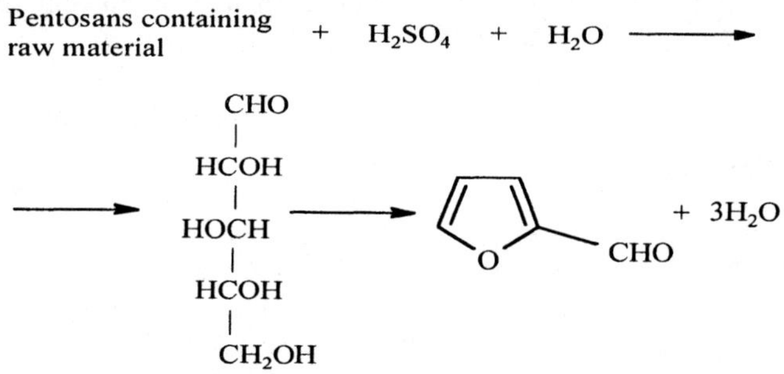

Figure 1.Furfural formation.

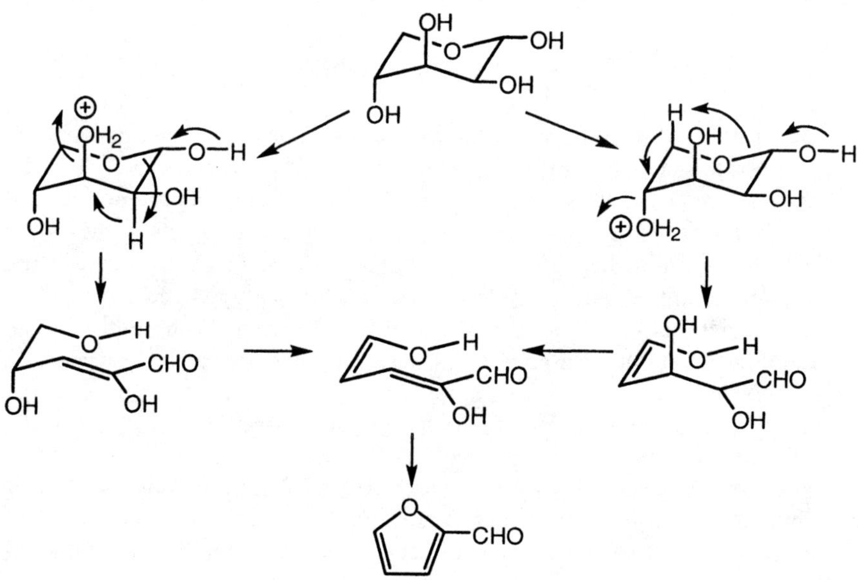

Figure 2. Furfural formation mechanism according to (4).

The rate of the first step of the furfural production process, a hydrolysis reaction, averages about 50 times higher than that of the second step, the dehydration reaction *(5)*. Hence, the furfural formation process limiting stage is the dehydration reaction. For a commercial-scale biomass conversion, the most important target is to maximize generation of pentoses and minimize a loss of cellulose. To achieve competitiveness with petrochemical processes, carbohydrate secondary degradation reactions must be eliminated and the furfural yield maximized. It is theoretically possible to obtain furfural as a monomer for organic synthesis and bioethanol as a motor fuel. However, in fact, up to now, production of these two chemical intermediates using dilute acid catalysis has been considered impossible because of close values of kinetic rate constants for furfural formation and cellulose degradation. This results in a 40-50% cellulose destruction during the furfural production process, and the low grade (chemically "inactive") residue may be used only as a fuel. Another problem, which has not been solved for 77 years of the industrial furfural production, was a yield of furfural not exceeding 55% from the theoretical one due to secondary reactions. The mechanism of the process had to be changed in order to solve both problems simultaneously. On the basis of a new non-traditional approach, namely, the theory of differentiated catalysis, the hydrolysis and dehydration reactions were performed simultaneously in one apparatus in industrial practice.

Furfural Production Using Small Amounts of Strong Catalysts

According to Vedernikov *(5),* the hydrolysis of pentosans and dehydration of pentoses in a one-step furfural production from a plant raw material in the presence of small amounts of concentrated catalyst solutions are accelerated differently by acetic and sulfuric acid (or other strong catalysts). Due to the marked chemosorption heat, the molecules of sulfuric acid sorbed on the surface of the particles are bound tightly to the polysaccharides by chemosorption and do not penetrate into the particles. The strength of the chemosorption between the molecules of sulfuric acid and polysaccharides increases with an increase in the initial acid concentration. It is minor for dilute solutions. As small amounts of sulfuric acid are used (2-4% of material mass), it covers less than 1% of the entire surface of particles.

Acetic acid, formed through treatment of wood with steam and uniformly distributed in the particles, catalyzes the hydrolysis of pentosans to monosaccharides. Further conversion of pentoses into furfural directly in the cell wall does not occur because acetic acid is too weak as a catalyst for dehydration at a low temperature. The pentoses formed in the process diffuse to the surface of particles where their dehydration to furfural occurs under the influence of sulfuric acid. Studies on the process kinetics *(6)* have confirmed that a higher concentration of sulfuric acid increases the rate of pentoses dehydration more markedly in comparison to hydrolysis of pentosans (Table I). This allows a decrease in the surplus of pentoses in the reaction system, reduces the possibility of side reactions and enhances the selectivity of the process.

Table I. Kinetic rate constant values as a function of catalyst concentration. k_1 - pentosan hydrolysis; k_2 - pentoses dehydration *(6)*

Concentration of H_2SO_4, %	$k_1 \cdot 10^2$, min^{-1}	$k_2 \cdot 10^2$, min^{-1}	k_1/k_2
10	14.58	0.251	58.1
20	15.76	0.816	19.3
30	16.50	1.024	16.1
60	15.84	1.450	10.9
90	16.17	1.954	8.3

The other avenue of furfural study with the target of improving the furfural production technology is a search for new catalysts. Up to now, this has been merely an empirical process. The catalytic activity is shown to be in a direct connection with the physico-chemical structure and properties of cations *(5,6)*. This phenomenon can be expressed by the functional relationship between the xylose dehydration rate constant and a value of the ionization potentials of an atom required for the formation of the corresponding cation (Figure 3). The dependence in the dehydration reaction rate constant upon the ionization potential relationship is similar to the Arrhenius ratio of reaction rate-temperature plot (Figure 4). An energetic interrelationship of these parameters is assumed. The obtained regularity of the catalytic activity of cations explains a certain progress in discovering the chemical mechanism of the process and provides a scientifically valid selection of catalysts as an alternative to empirical approaches.

To realize the furfural production process *(7)*, an original two-shaft helix shaped blade continuous mixer and an air-disperser has been developed (Figure 5). When modeling the mixing processes of the raw material with a catalyst, diffusional model parameters of the hydrodynamic structure of the material internal flows, depending on the length and equivalent diameter of the mixer, have been determined. The optimum layout diagram of the blades on the mixer shafts, their configuration, shaft distance as well as the number of their revolutions have been experimentally determined. The optimum combination of these parameters has provided a uniform distribution of the catalyst in the raw material mass. The manufacturing of the equipment (of various capacities) for raw material mixing with catalyst solutions is currently under way.

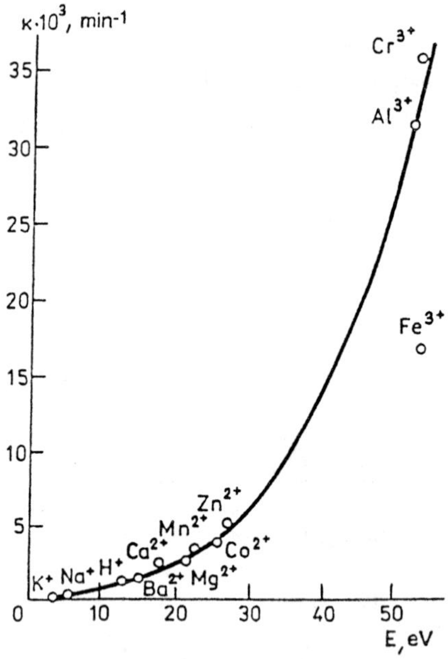

Figure 3. Xylose dehydration rate constant as a function of catalyst cation ionization potential.

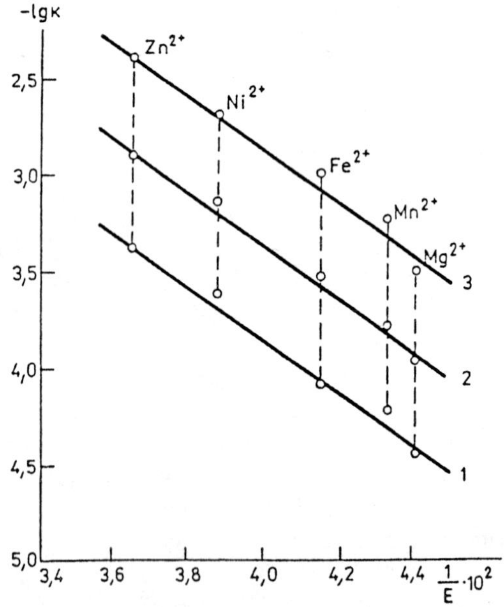

Figure 4. Furfural formation from birch wood: relationship between the dehydration reaction rate constant and the catalyst ionization potential at different temperatures (1 – 405 K; 2 – 415 K; 3 – 425 K).

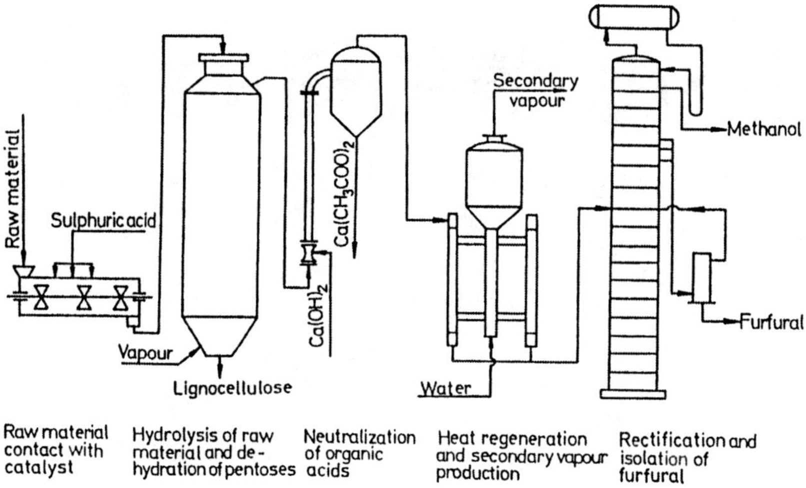

Figure 5. Flow diagram of furfural production using small amounts of strong catalysts.

Integration of Furfural Production with Other Biomass Treatment Methods

The integration of various production systems allows an increase in the diversity of value-added products, increased competitiveness and a solution to environmental problems. In the best case, such an integration eliminates wastes *(8,9)* because the waste from one production system can become a raw material for another one (Zero Emissions integrated cluster, United Nations University *(1,8)*).

The production of ethanol alone is not economically feasible without continued subsidies but a plant becomes profitable if it produces ethanol and furfural as co-products *(10)*. The technology development in the Latvian State Institute of Wood Chemistry (LSIWC) provides an excellent possibility for producing both furfural and ethanol. As a result, the problem of the complete utilization of the deciduous wood polysaccharide complex yielding furfural and fermentable sugars to be used subsequently for the production of bioethanol and other microbial synthesis products has been solved. Residual lignin could be used as a calorific fuel.

Another example of integration is the production of levoglucosan obtained as a result of furfural production from the residual cellulosic part by fast pyrolysis. The schematic outline of lignocellulose processing is shown in Figure 6.

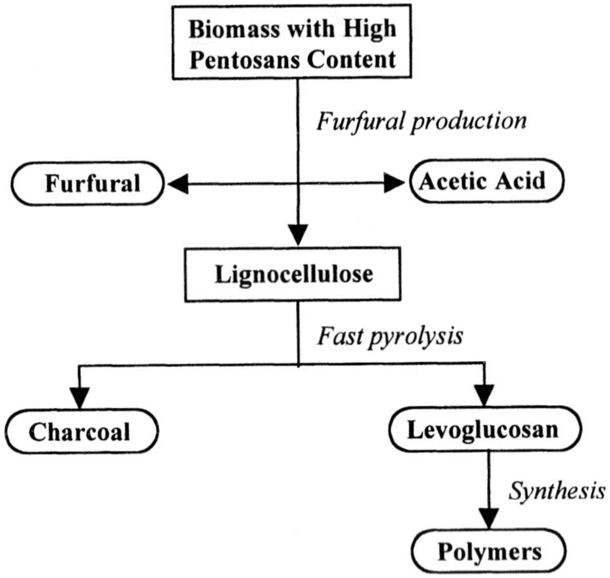

Figure 6. Integration of concentrated acid catalyzed furfural production technology with fast pyrolysis treatment.

One of products of cellulose thermolysis is 1,6-anhydro-β-D-glucopyranose, levoglucosan. Upon vacuum thermolysis, its yield exceeds 70% from the weight of cellulose. This carbohydrate anhydride is of practical interest as a reactive triol. Hence, the mechanism of its formation during thermolysis of lignocellulose and the technological processes had been studied *(11)*. It was revealed in the late 1960s that levoglucosan is formed with a good yield during fast pyrolysis of hardwood sawdust and other fine particle cellulosic materials. For the first time, the pilot-plant scale fast pyrolysis of lignocellulose to obtain levoglucosan was carried out at LSIWC. A good yield of levoglucosan (20 to 26% from the weight of o.d. lignocellulose of concentrated sulfuric acid catalysed furfural production, or 47.5 to 63.0% from the weight of cellulose) was demonstrated under these conditions. In the case of thin particles, the rate of thermal destruction is determined by the speed of consecutive chemical reactions, which proceed in two stages: at the first stage, a dramatic decrease in the degree of polymerization of cellulose chains, and then two competing reactions, i.e., a low-temperature dehydration reaction with the formation of anhydrocellulose, followed by a high-temperature depolymerization reaction with the formation of levoglucosan. The high temperature reaction of the depolymerization of a cellulose macromolecule with the formation of levoglucosan probably proceeds via formation of intermediate compounds. Fast thermolysis of lignocellulosic materials in a flow of superheated steam was realized in a cyclone type thermoreactor with a production capacity of 30 kg of raw material per hour. The steam jacket prevents losses of heat and, in addition, warms up the decomposed material. The removal of the mix of volatile products of thermolysis and heat-carrier is carried out through the central channel.

One of the basic problems in introducing the production of levoglucosan is its isolation from the products of thermodestruction of lignocellulose and obtaining it in the crystalline form (>95% purity). Technical levoglucosan was obtained by the acetone method consuming 2.5 ml of acetone per 5 g of thermolysis tar (evaporated condensate) with a moisture content of no more than 10% containing 45 to 70% of levoglucosan (from the weight of o.d. pitch). The research carried out at LSIWC on the process of levoglucosan crystallization from a mixture of thermolysis tar and acetone has shown that crystallization takes no more than 2 hours, since 54% of levoglucosan (from its content in the thermolysis tar) is obtained from the mixture as technical levoglucosan in 60 minutes and at the maximal yield 55.6% in 105 minutes. The acetone method of levoglucosan isolation is simple and provides a product containing more than 90% levoglucosan. It was shown that technical levoglucosan purified by the selective extraction method crystallizes yielding a product containing 95 to 97% of the basic substance.

Due to the presence of a 1,6-anhydro cycle capable of cleavage, as well as three secondary hydroxyl groups in levoglucosan, the latter can be used for the synthesis of various low- and high-molecular compounds (Figure 7). Levoglucosan derivatives, synthesized according to reactive hydroxyl groups, have been used in a variety of applications. Polyethers may be used in the production of rigid polyurethane foams. The introduction of levoglucosan and its esters/ethers in a polyurethane foams composition enhances the heat stability and rigidity of polyurethane foams, while the use of Cl- and F-containing ethers inhibits the flammability. Levoglucosan trimethacrylate is an active cross-linking monomer, which can be used in the production of high-molecular rubbers applicable in the manufacture of the accessory parts of dynamic connection seals. Oligocarbonate methacrylates are offered and tested as components of anaerobic materials, photopolymerizing compositions for galvanoplastic models and optical disc carriers. Levoglucosan epoxy resins are tested as heat setting additives in polyvinyl chloride compositions. However, to our knowledge no commercial products based on levoglucosan have yet been marketed *(12)*.

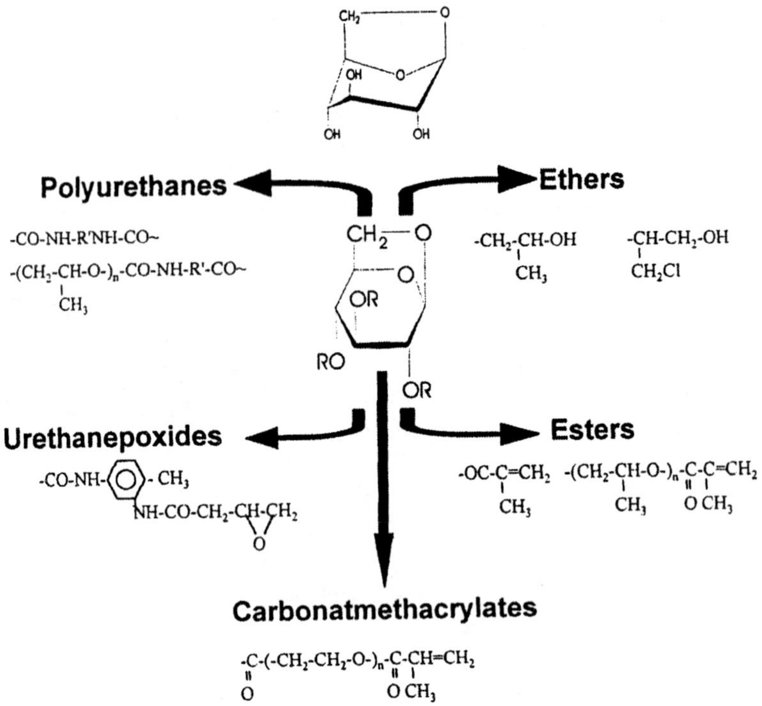

Figure 7. Levoglucosan-based polymer products synthesized at LSIWC

The third attempt to integrate furfural production with other biomass processing systems involves a steam explosion biomass treatment after furfural separation *(13)*. This variant also emphasizes the application of residual lignin as a value-added product. Steam explosion is chosen to produce pulp which can be separated into microcrystalline cellulose, lignin and reducing substances. A combination of the latter two with the untreated portion of biomass (fiber materials) allows fiberboards and composite materials (plastics) to be obtained. The carbonization of plastics at high temperatures gives a carbon material (wood ceramics), which acts as a concentrated carbon sink, while volatile products of pyrolysis (producer gases) formed during the process are passed to a boiler house. The flow diagram which includes also the above mentioned integration possibilities is shown in Figure 8. This integrated biomass processing approach provides a complete wood and non-wood plant materials refinery into value-added chemicals, materials and biofuels. As for now, some blocks depicted in the flow diagram are indeed realized to produce market chemicals (furfural, ethanol, etc.). Some are in the commercialization phase (microcrystalline cellulose, wood plastics, wood ceramics, etc.), and some need future improvement (levoglucosan). The proposed Zero Emissions Refinery Cluster *(8,9)* is energetically self-sufficient and environmentally sound.

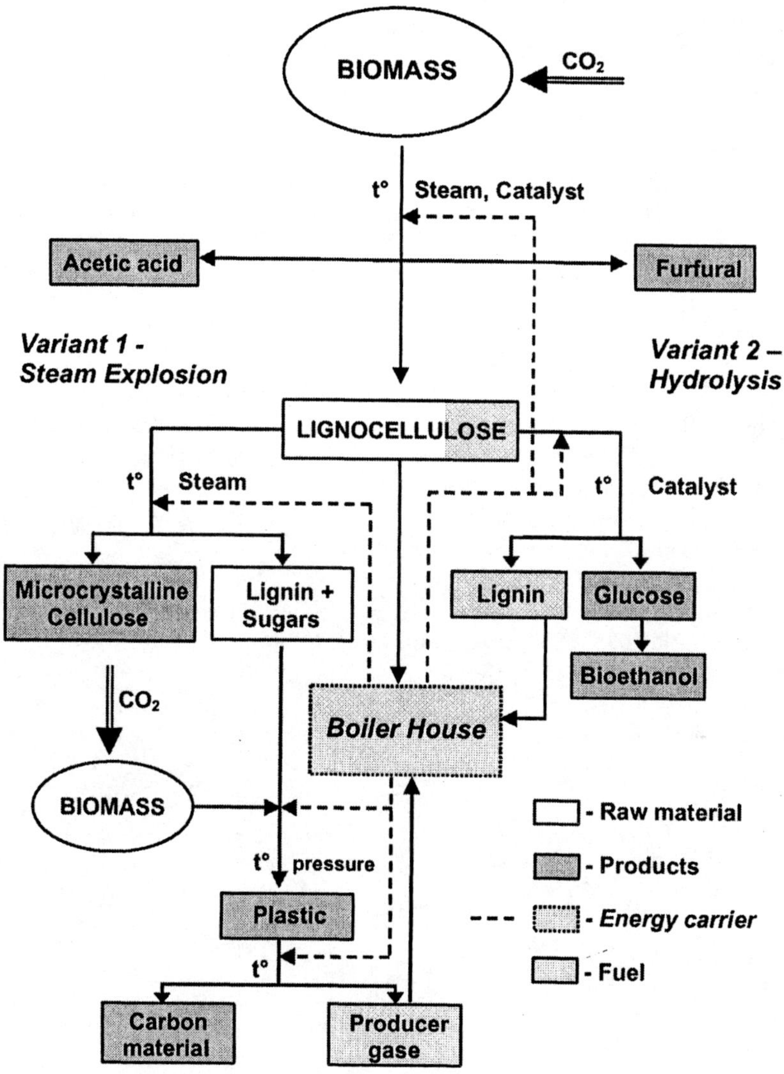

Figure 8. Zero Emissions Biomass Refinery Cluster.

References

1. Heden, C.-G.; Gravitis, J. *Proc. Intern. Symp. Towards Zero Emissions: The Challenge for Hydrocarbons,* Rome, Italy, March 11-13, **1999**, 46-52.
2. Wegener, G. *Meeting of Intern. Expert Council on the Chemistry of Vegetal Resources (COCVER),* **1995**, 50-78.
3. *Chemical Economics Handbook.* SRI International, **1991**.
4. Garrett, E. R.; Dvorchik, B.H. *J. Pharmaceut. Sci.* **1969**, 58, 813-820.
5. Vedernikov, N. *Hemicelluloses;* Zinatne: Riga, **1991**, 209-222 (in Russian).
6. Vedernikov, N. *Proc. 10th Intern. Symp. Wood and Pulping Chemistry,* Yokohama, Japan, June 7-10, **1999**, 3, 468-470.
7. Vedernikov, N. et al. Russian pat. 1365674, **1994**.
8. Gravitis, J.; Della Senta, T.; Williams, E. D. *Proc. Symp. Biomass Conversion,* Sapporo, Japan, October 26, **1997**, 1-18.
9. Gravitis, J.; Suzuki, M. *Proc. IV Intern. Congress on Energy, Environment and Technological Innovation,* Rome, Italy, September 20-24, **1999**, 1, 695-700.
10. Van Dyne, D. L.; Blase, M. G. *Proc. 4th Biomass Conf. Americas,* Oakland, USA, August 29-September 2, **1999**, 1, 363-368.
11. Pernikis, R.; Zandersons, J.; Lazdina B. *Developments in Thermochemical Conversion.* Bridgwater, A.V.; Boocock, D.G.B. Eds.; Blackie Academic & Professional: London, **1997**, 3, 536-548.
12. Longley, C. J.; Fung, D. P. C. *Papers of the Intern. Conf. Advances in Thermochemical Biomass Conversion,* **1994**, 2, 1484-1494.
13. Kokorevics, A.; Gravitis, J.; Bikovens, O. *13th Intern. Congress of Chemical and Process Engineering, CHISA'98,* Praha, Czech Republic, August 23-28, **1998**, CD-ROM of Full Texts, Report 5.65, 1-20.

Chapter 10

Pretreatment Processes to Increase Pyrolytic Yield of Levoglucosan from Herbaceous Feedstocks

Robert C. Brown[1], Desmond Radlein[2], and Jan Piskorz[2]

[1]Center for Sustainable Environmental Technologies, Iowa State University, Ames, IA 50011
[2]Resource Transforms International, Ltd., 110 Baffin Place, Unit 5, Waterloo, Ontario N2V 1Z7, Canada

We are investigating various pretreatment processes to increase the pyrolytic yield of levoglucosan from herbaceous biomass. During fast pyrolysis of biomass, alkali and alkaline earth metals occuring naturally in biomass serve as catalysts to degrade lignocellulose to char. If these cations are removed from the biomass before pyrolysis, lignocellulose is depolymerized to anhydrosugars with very high yields. Previous studies of the effect of demineralization on yields of levoglucosan focused on woody feedstocks. We are evaluating pretreatment processes to remove alkali from herbaceous feedstocks such as corn stover and switchgrass. Three pretreatments of herbaceous biomass were evaluated: acid hydrolysis, washing in dilute nitric acid, and washing in dilute nitric acid with the addition of $(NH_4)_2SO_4$ as a pyrolytic catalyst. Only the water wash proved ineffective. However, demineralization using dilute nitric acid offers the best combination of simplicity and effectiveness.

Introduction

We are investigating various pretreatment processes to increase the pyrolytic yield of levoglucosan from herbaceous biomass. This work is motivated by our effort to convert lignocellulosic material into simple sugars suitable for fermentation. In particular, we are investigating thermal depolymerization of corn stover and switchgrass into anhydrosugars and their subsequent hydrolysis and fermentation to lactic acid.

Fast pyrolysis combined with fermentation offers the prospect of circumventing the slow and costly acid and enzymatic hydrolysis of lignocellulose that currently stand in the way of converting low-value lignocellulose into high-value chemical products. Among the challenges to this approach for producing chemicals and fuels from crops and plant materials is maximizing the yield of levoglucosan and other anhydrosugars. This study examines various pretreatments of biomass feedstocks that can achieve this end.

Background

Lignocellulosic plant materials are underutilized by the agricultural processing industry. Indeed, lignocellulose, in the form of oat hulls, corn stover, wheat straw, and similar materials are usually considered as wastes. Corn stover is by far the largest lignocellulosic feedstock available in the midwestern United States. Iowa alone produces 10.5 million tons/yr of corn residue *(1)*. It has long been recognized that cellulose can be converted to sugars followed by fermentation to alcohol or organic acids *(2)*. If agricultural wastes such as corn stover could be economically converted to industrial chemicals, it would represent an important new development for agriculture.

Considerable research has been directed at converting crops and plants into industrial chemicals. The major stumbling block in this effort has been fractionation of lignocellulose into its constituents, lignin and cellulose, and hemicellulose. Research in the past has focused on acid pretreatments and enzymatic hydrolysis, which have proven too costly to date for commercial applications.

Fast pyrolysis offers a possible solution to the problem of economical fractionation. Pyrolysis is the thermal decomposition of carbon-bearing materials in the absence of oxygen to generate gaseous, liquid, and solid products *(3)*. Fast pyrolysis, a special case of pyrolysis, is a non-equilibrium process that occurs under very specific conditions of moderate temperature (450-600°C) and short residence time (on the order of 1 s). Reaction is followed by rapid quenching, which prevents high molecular weight compounds from decomposing to low-molecular weight gases *(4)*. Liquid yields as high as 70% have been produced under fast pyrolysis conditions *(5)*.

Fast pyrolysis liquid is a complicated mixture of organic compounds arising from the uncontrolled degradation of cellulose and lignin in the biomass. The liquid is highly oxygenated, approximating the elemental composition of the feedstock. It contains many different compounds, including phenols, sugars, and both aliphatic and aromatic hydrocarbons *(6)*. Research is ongoing to make these liquids compatible with petrochemical processing *(7)*. However, another possibility is to change process conditions to generate a liquid that is compatible with biochemical processing.

Scott and coworkers *(8)* at the University of Waterloo in Ontario, Canada, have proposed an alternative route for "cracking" lignocellulose based on fast pyrolysis to produce fermentable sugars. They recognized that alkali and alkali earth metals in the biomass serve as catalysts that degrade lignocellulose to gas and char. If these cations are removed from the biomass before pyrolysis, the lignocellulose is depolymerized to anhydrosugars with very high yields. In other words, the biomass is delignified.

In tests with poplar wood, Scott and his colleagues observed several important changes in the composition of pyrolysis oil as a result of demineralizing the wood before pyrolysis *(9)*. First, the yield of organic liquids increased from 66% to 80%. Second, the yield of simple fermentable sugars more than doubled, although the amount was still modest. However, the amounts of anhydrosugars increased several fold to commercially significant levels. In particular, levoglucosan increased from less than 3% to over 30% of the total organic compounds in the liquid. Hydrolysis of levoglucosan occurs readily in hot dilute acid to yield glucose. The resulting sugar solution has been successfully fermented *(9)*.

So and Brown *(10)* have recently compared the economics of fast pyrolysis integrated with a fermentation step to two other lignocellulose-to-ethanol conversion technologies: simultaneous saccharification and fermentation and dilute sulfuric acid hydrolysis and fermentation. The three technologies were compared in terms of capital costs, operating costs, and ethanol production costs. The analysis showed that fast pyrolysis integrated with a fermentation step is comparable with the other two processes.

Most previous work on production of high sugar pyrolysates has used wood as feedstock, which differs in many respects from the herbaceous material of interest to most agricultural operations. For example, herbaceous materials contain more potassium than woody materials *(11)*. Since the catalytic activity of potassium under conditions of pyrolysis are considerably greater than for calcium *(12)*, it may be more difficult to thermally depolyermerize herbaceous plant materials. However, potassium is also more water soluble than calcium and thus easier to remove. This study investigates various pretreatment options for their ability to increase pyrolytic yields of levoglucosan from herbaceous feedstocks.

Experimental Equipment and Procedures

Three pretreatments were evaluated for their effectiveness in enhancing pyrolytic yields of levoglucosan from corn stover and switchgrass. These are summarized in Table I.

Table I. Pretreatment options tested for production of pyrolysis syrups

Pretreatment option	*Procedure*
Acid hydrolysis	Soak chopped biomass in 5.0 wt % H_2SO_4 at 100 C for 2 hours; rinse in distilled water
Demineralization	Wash chopped biomass in 0.25 wt % nitric acid at 25 C
Demineralization with pyrolysis catalyst	Same as demineralization treatment except that $(NH_4)_2SO_4$ is added afterward as catalyst (addition rate of 0.10 wt %)

Pretreatment of biomass was preceded in all cases by air drying to less than 10% moisture and chopping to 0.5 mm long fibers or flakes. In this form the material is referred to as "raw" feedstock. Also in common for the three pretreatments was the use of a 5.9 cm dia. glass column of 58 cm height to wash or rinse the biomass, as detailed below.

As the name implies, the acid hydrolysis pretreatment is intended to partially hydrolyze the lignocellulosic feedstock as well as remove alkali from it. This process converts hemicellulose to pentose sugar as well as removes alkali. Biomass was added to a 5 wt % solution of sulfuric acid in the proportion of about 150 g biomass for every liter of solution and boiled for 2 hours. The mixture was transferred to the glass column and excess liquid drained from the bottom. At this point 2 L of distilled water was pumped through the column at a superficial velocity of 0.08 mm/s to rinse acid from the biomass. The pretreated biomass was then dried at 105°C for subsequent pyrolysis.

Demineralization attempts to dissolve alkali from the biomass using a weak acid solution but without hydrolyzing the lignocellulose. In the case of the switchgrass, approximately 5 L of distilled water was first washed through the column packed with 200 g of switchgrass before washing with 2.7 L of the 0.25 wt% nitric acid. This water prewash was deemed unnecessary in the cornstover trial; thus 200 g of cornstover was slurried in 2 L of 0.25 wt% nitric acid solution and charged to the column, allowing excess acid to drain off. At this point 1.5 L of additional nitric acid solution was pumped through the column to remove additional alkali from the biomass. Both switchgrass and cornstover were dried at 105°C for subsequent pyrolysis.

A portion of the demineralized (nitric acid treated) feedstock was used to test the effect of ammonium sulfate ($(NH_4)_2SO_4$) on pyrolytic sugar yield. Shafizadeh *(13)* has postulated a catalytic effect by acid during pyrolysis. Scott et al. *(8)* have suggested the addition of $(NH_4)_2SO_4$ to demineralized feedstock as a way of adding acid catalyst. Presumably the sulfate decomposes to ammonia and sulfuric acid, providing a controlled release of acid to the pyrolyzing feedstock. For the present test, food grade $(NH_4)_2SO_4$ was mixed with sufficient water to thoroughly wet the feedstock, which was then dried with constant turning to assure uniform distribution. An addition rate of 0.10 wt% was tested on demineralized samples of both switchgrass and cornstover.

Raw biomass and all the pretreated samples except those receiving the water wash (for reasons explained in the Results and Discussions section) were thermally processed in the fast pyrolysis pilot plant at RTI, Ltd. This bubbling, fluidized bed reactor operates between 360°C and 490°C with gas residence times on the order of 2 s. Details on this system can be found in Reference 14. Operating conditions for pyrolysis of the switchgrass and cornstover samples are detailed in Tables II and III, respectively.

Table II. Operating Conditions for Fast Pyrolysis of Switchgrass

Operating Variable	*Raw*	*Acid Hydrolysis*	*Demineralized*	*Demineralized + catalyst*
Moisture, wt%	8.7	12	3.6	7.0
Ash, wt%	5.1	4.5	3.1	3.2
Temperature, C	449	435	449	449
Feed rate, g/h	37.8	34.4	47.2	35.2

Table III. Operating Conditions for Fast Pyrolysis of Cornstover

Operating Variable	*Raw*	*Acid Hydrolysis*	*Demineralized*	*Demineralized + catalyst*
Moisture, wt%	6.8	2.2	2.2	4.5
Ash, wt%	5.9	5.0	4.2	4.3
Temperature, C	450	430	440	430
Feed rate, g/h	71	100	88	110

Water analysis of liquids was by Karl Fischer titration. Liquids were analyzed by GC for qualitative or semi-quantitative characterization. Quantitative analysis of selected liquid components was performed by HPLC. The sample was prepared by diluting it with two parts water to cause phase separation of "pyrolytic lignin." Simple sugars, anhydrosugars, and other lower molecular weight components such as acetic acid and hydroxyacetaldehyde remain in the aqueous phase.

Note from these tables that all pretreatments significantly reduce the ash content of the raw biomass on a weight percentage basis (12 to 37%). The ash that remains is insoluble in acid and is assumed to be mainly silica. Acid hydrolysis shows the smallest reduction in ash content of the three pretreatments. In fact, acid hydrolysis is probably very efficient in removing alkali but it also removes carbohydrate from the feedstock in the form of pentose sugar hydrolyzed from hemicellulose. In this case, the percentage change in ash content does not directly indicate the loss in alkali. This loss of hemicellulose from the acid hydrolyzed samples also explains several phenomena observed in the analysis of the pyrolysis liquid, as subsequently described.

Results & Discussion

The yield of char, water, organics, and gases is compared in Table IV for pyrolysis of raw switchgrass and switchgrass pretreated by acid hydrolysis, demineralization, and demineralization with added catalyst. Table V is the corresponding data for pyrolysis of cornstover. From these data it is clear that all three pretreatments increase the yield of organic liquid (14 - 37%) and decreased the yield of gases, which is highly desirable for the proposed process. The enhancement in the yield of organics was especially pronounced for switchgrass. Untreated switchgrass produced 52.2% organics compared to 59.1% organics for pyrolysis of untreated cornstover. However, organic yields were comparable for the two materials after pretreatment.

The yield of organic compounds in pyrolysate derived from switchgrass and cornstover are given in Tables VI and VII, respectively. The principal anhydrosugars are levoglucosan and cellobiosan, a dimeric form of levoglucosan. The total concentration of these two compounds in the organic fraction of raw switchgrass pyrolosate was only 4.4%. This value increased dramatically to 39% for hydrolyzed switchgrass, 22% for demineralized switchgrass, and 27% for

Table IV. Pyrolysis yields for pretreatments of switchgrass

Constituent Wt% maf	*No Pretreatment*	*Acid Hydrolysis*	*Demineral-ization*	*Demineral-ization + catalyst*
Char	14.5	17.6	11.0	10.7
Water	9.65	8.52	11.7	11.4
Organics	52.2	62.1	65.7	67.8
Gases	23.7	11.8	11.7	10.1

Table V. Pyrolysis yields for pretreatments of cornstover

Constituent Wt% maf	*No Pretreatment*	*Acid Hydrolysis*	*Demineral-ization*	*Demineral-ization + catalyst*
Char	15.8	15.9	13.2	13.23
Water	2.57	7.96	10.6	10.39
Organics	59.1	67.7	67.2	68.5
Gases	22.6	8.44	9.02	7.88

Table VI. Yields of organic compounds in pyrolysis syrups for pretreatments of switchgrass

Organics Wt %	*No Pretreatment*	*Acid Hydrolysis*	*Demineral-ization*	*Demineral-ization with catalyst*
Cellobiosan	trace	3.27	5.65	6.53
Levoglucosan	4.42	35.62	16.10	20.65
Hydroxy-acetaldehyde	13.51	2.49	7.19	7.58
Formic acid	5.75	trace	4.85	4.02
Acetic acid	4.24	0.41	3.07	2.59
Acetol	4.46	-	-	-
Formaldehyde	2.92	0.54	-	2.20
Pyrolytic lignin	30.39	22.92	16.11	13.61

Table VII. Yields of organic compounds in pyrolysis syrups for pretreatments of cornstover

Organics Wt %	*No Pretreatment*	*Acid Hydrolysis*	*Demineral-ization*	*Demineral-ization with catalyst*
Cellobiosan	trace	4.97	4.55	3.34
Levoglucosan	2.75	23.10	17.69	20.12
Hydroxy-acetaldehyde	11.57	3.93	5.97	3.73
Formic acid	2.61	0.73	trace	trace
Acetic acid	3.40	0.40	1.51	1.26
Acetol	4.53	trace	trace	trace
Formaldehyde	2.75	0.70	1.63	trace
Pyrolytic lignin	33.40	20.08	16.89	17.74

demineralized switchgrass with ammonia sulfate catalyst added. Similarly, the total concentration of anhydrosugar in the organic fraction of raw cornstover pyrolysolate was only 2.8% increasing to 28% for hydrolyzed cornstover, 22% for demineralized cornstover, and 23% for demineralized cornstover with ammonia sulfate catalyst added.

Normally hydroxyacetaldehyde, which is formed in competition with levoglucosan during cellulose pyrolysis *(15)*, is catalyzed by the presence of cations such as alkali metal. Indeed, our results show that hydroxyacetaldehyde declines as levoglucosan increases in pyrolysate. However, hydroxyacetaldehyde remains a significant constituent of the pyrolysate from pretreated biomass, which suggests that some alkali remains in the feedstock even after pretreatment.

Another explanation is the formation of hydroxyacetaldehyde from thermally unstable hemicellulose in the samples. However, levoglucosan yield is strongly correlated to hydroxyacetaldehyde for all pyrolysates, including those obtained from biomass receiving acid hydrolysis pretreatment. Since the hydrolyzed samples are substantially depleted of hemicellulose, it seems unlikely that the hydroxyacetaldehyde is derived from hemicellulose.

Demineralization with the addition of $(NH_4)_2SO_4$ as a catalyst increased total yield of anhydrosugars (%wt organics in pyrolysate x %wt anhydrosugars in organics) from switchgrass by about 30% and from cornstover by about 8% compared to demineralization alone. The reasons for these differences between switchgrass and cornstover are not clear but are probably related to the higher hemicellulose content of switchgrass. Additional tests are required to justify the use of $(NH_4)_2SO_4$ as a catalyst in the proposed process.

The acid hydrolysis pretreatment yielded the highest percentages of anhydrosugars. However, hydrolysis removes hemicellulose from biomass, thus enriching the sample with cellulose at the start of pyrolysis. Accordingly, higher percentages of anhydrosugars should be expected in pyrolysate from hydrolyzed biomass. Indeed, based on the summative analysis of typical switchgrass, we might expect hydrolysis to yield 42% more levoglucosan than does demineralization while hydrolysis of relatively hemicellulose-poor cornstover would yield 17% more levoglucosan than the demineralization pretreatment. These predictions are consistent with the experimental data: for switchgrass, hydrolysis yielded 41 – 72% more levoglucosan than did demineralization while the enhancement was 27% for cornstover. The enrichment of lignin in the hydrolyzed samples is also consistent with this argument.

This result does not necessarily make hydrolysis more attractive than demineralization; effective measures for recovering pentose sugar from the hydrolysis waste stream are required. Our analysis suggests that there is no clear advantage in the acid hydrolysis pretreatment, which is more complicated and expensive than the demineralization pretreatment. Furthermore, the hydrolyzed biomass is sticky and hard to feed into the pyrolysis reactor.

Conclusions

Both demineralization with dilute nitric acid and hydrolysis with hot sulfuric acid appeared effective in removing alkali from herbaceous biomass and dramatically increasing the pyrolytic yield of anhydrosugars (levoglucosan and cellobiosan). Ammonium sulfate showed catalytic activity ranging from 5 – 30%; additional testing is required to justify its use in the proposed process. Although acid hydrolysis produced the highest percentage yields of anhydrosugars, this is expected for the hemicellulose-depleted material resulting from this pretreatment.

Hydroxyacetaldehyde, which forms in competition with levoglucosan during pyrolysis of cellulose, remains a significant constituent of pyrolysate from pretreated biomass. Further tests are required to determine whether this is the result of residual alkali in the pretreated biomass.

Acknowledgements

We appreciate the assistance of James Pollard in preparation of feedstocks used in this study. We gratefully acknowledge the financial support of the Iowa Energy Center, the U. S. Department of Energy's Great Lakes Regional Biomass Energy Program, and the United States Department of Agriculture in performing this study.

References

1. Hesel, Z. R.; Wedin, W. F. In *Energy in Agriculture*; Elsevier Science Publishers: Amsterdam, 1983.
2. Grohmann, K.; Wyman, C. E.; Himmel, M. E. In *Emerging Technologies for Materials and Chemicals from Biomass*, Rowell, R. M; Schultz, T. P.; R. Narayan, Eds.; ACS Symp. Series 476; American Chemical Society: Washington, D.C., 1992; pp. 354-393.
3. Bridgwater, A. V.; Bridge, S. A. In *Biomass Pyrolysis Liquids Upgrading and Utilization*; Bridgwater, A. V.; Grassi, G., Eds.; Elsevier Science: NY, 1991; pp. 11-92.
4. Graham, R. G.; Bergougnou, M. A.; Mok, L. K. S.; De Lass, H. I, In *Fundamentals of Thermochemical Biomass Conversion*, Overend, R. P.; Milne, T. A.; Mudge, L. K., Eds.; Elsevier Applied Science Publishers: London, 1985.
5. Hayes, R. D., ACS Symp. Ser. 376; American Chemical Society: Washington, D.C., 1988; pp. 8-15.
6. Desbene, P. L.; Essayegh, M.; Desmazieres, B.; Basselier, J. J. In *Biomass Pyrolysis Liquids Upgrading and Utilization*; Bridgwater A. V.; Grassi, G., Eds.; Elsevier: NY, 1991; pp. 155-176.
7. Rupp, M. In *Biomass Pyrolysis Liquids Upgrading and Utilization*; Bridgwater, A. V.; Grassi, G., Eds.; Elsevier Science: NY, 1991; pp. 219-241.
8. Scott, D. S.; Piskorz, J.; Radlein, D.; Czernik, S. In *Pyrolysis and Gasification*; Fenero, G. L.; Maniatis, K.; Buekens, A.; Bridgwater, A. V.; Eds. Elsevier Applied Sci. Publ., London, 1989; pp. 201-208.
9. Piskorz, J.; Majerski, P.; Radlein, D.; Scott, D. S.; Landriault, Y. P.; Notarfonzo, R. P.; Vijh, D. K. In *Proceedings of the Third Biomass Conference of the Americas*; Overend, R. P.; Chornet, E. Eds.; Pergamon: Montreal, Ontario, 1997, pp. 823-833.
10. So, K.; Brown, R.C. *Appl. Biochem. Biotechnol.* **1999**, *77-79*, 633-640.
11. Baxter L.; Miles, Sr., T.; Miles, Jr., T.; Jenkins, B.; Dayton, D.; Bryers, R.; Olden, L. *Alkali Deposits Found in Biomass Boilers*; Vol. II, NREL/TR433-8142, SAND96-8225, 1996.

12. Lang, R. J.; Neavel, R. C. *Fuel* **1982**, *61*, 620.
13. Shafizadeh, F.; Stevenson, T. T. *J. Appl. Polymer Sci.* **1982**, *27*, 4577 – 4585.
14. Scott, D. S.; Majerski, R.; Piskorz, J.; Radlein, D. *J. Anal. Applied Pyrol.* **1999**, *51*, 23-37.
15. Richard, G. N. *J. Anal. Appl. Pyrol.* **1987**, *10*, 251-256.

Chapter 11

3-Dehydroshikimic Acid: A Building Block for Chemical Synthesis from Renewable Feedstocks

K. M. Draths[1], Spiros Kambourakis[1], Kai Li[1], and J. W. Frost[1,2,*]

Departments of [1]Chemistry and [2]Chemical Engineering, Michigan State University, East Lansing, MI 48824

3-Dehydroshikimic acid is an ideal building block for chemical synthesis from feedstocks derived from renewable resources. Microbial catalysts that synthesize 3-dehydroshikimic acid from starch-derived glucose or glucose:xylose:arabinose mixtures that mimic corn fiber hydrolysate achieve high titers of product in high yield. Catalytic conditions ranging from microbial catalysis to traditional chemical catalysis have been developed that transform 3-dehydroshikimic acid into protocatechuic acid, vanillin, catechol, adipic acid, gallic acid, and pyrogallol.

Chemical synthesis has traditionally relied on a building block strategy where relatively few chemicals are identified as primary building blocks that are subsequently transformed into secondary building blocks and other materials of industrial interest. Of the top 50 large volume chemicals produced annually in the United States, the 28 organic compounds on the list are derived from only 8 primary building blocks (*1*). One consequence of the building block approach is that the vast majority of industrial chemicals are synthesized from a limited pool of feedstocks. Of the approximately 170 compounds produced annually in the U.S. in volumes exceeding 4.5 million kilograms, 98% are derived from petroleum and natural gas (*1*). Given that both these feedstocks are nonrenewable resources, the ability to synthesize the bulk of industrial chemicals is not limitless. Successful transfer of some chemical synthesis from nonrenewable feedstocks to those derived from renewable resources requires identification of new building blocks to emulate this strategy. Ideal building blocks must be easily synthesized in high yield from inexpensive, readily available renewable feedstocks and must be converted through various catalytic transformations into a variety of materials of industrial interest.

3-Dehydroshikimic acid (DHS) is presented as a useful chemical building block. Microbial catalysts have been created that convert glucose or mixtures of glucose, xylose, and arabinose into DHS in high yield. Glucose is currently obtained from starch, although cellulose may ultimately be a less expensive source, whereas glucose, xylose, and arabinose mixtures model the feedstock that may eventually be derived from inexpensive corn fiber. Optimization of microbial catalysts that utilize inexpensive feedstocks that might become available in the future is also possible. Using combinations of microbial catalysis and traditional chemical catalysis, DHS is transformed into a spectrum of chemicals of industrial interest (Figure 1).

Microbe-Catalyzed Synthesis of DHS

3-Dehydroshikimic acid is a hydroaromatic intermediate of the aromatic amino acid biosynthetic pathway (Figure 2) (*2*). Existing in microorganisms and plants, this pathway gives rise to the aromatic amino acids as well as several aromatic vitamins. As an intermediate of biosynthesis, quantities of DHS found in natural sources are extremely limited. Development of a process for large-scale DHS synthesis may lead to interest in DHS in its own right. For example, DHS was recently found to be a potent antioxidant (*3*). Used in commercial preparation of foods, materials, medicinals, and cosmetics, antioxidants are designed to interfere with oxidative decomposition processes. When DHS was tested for its ability to inhibit peroxide and hydroperoxide formation in lard, DHS provided complete protection against oxidation for a period of 28 days and was equal to or superior to the antioxidant activities of propyl gallate, gallic acid, and TBHQ (*3*), all of which are used as commercial antioxidants. Furthermore, as a small chiral molecule possessing several stereocenters and useful functionality, DHS might be an attractive synthon in the pharmaceutical arena if it were available in sufficient quantities.

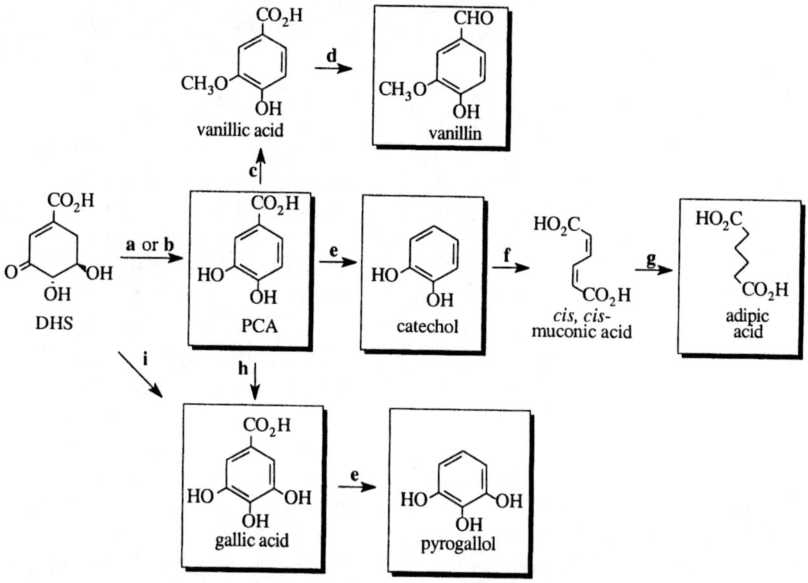

Figure 1. DHS as a chemical building block. Boxed structures are of current industrial value. Catalysts: (a) DHS dehydratase (aroZ); (b) reflux; (c) catechol-O-methyltransferase (comt); (d) aryl-aldehyde dehydrogenase; (e) PCA decarboxylase (aroY); (f) catechol 1,2-dioxygenase (catA); (g) Pt/C, H_2; (h) p-hydroxybenzoate hydroxylase (pobA); (i)$Cu(OAc)_2$, HOAc.*

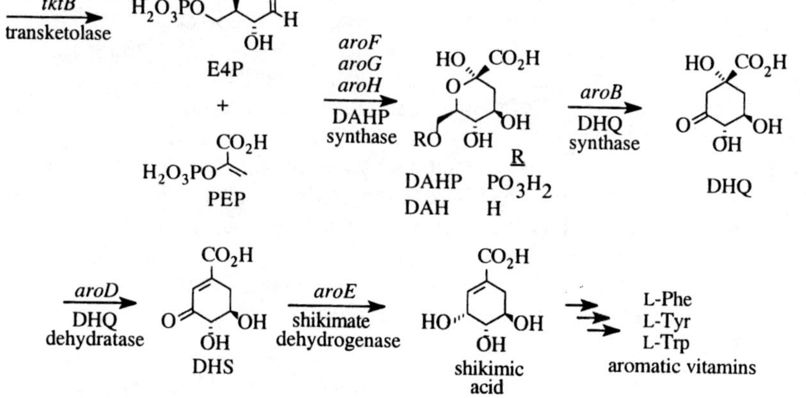

Figure 2. The aromatic amino acid biosynthetic pathway.

Synthesis of DHS from Glucose

Creation of microbial catalysts for high-yielding synthesis of DHS from glucose began with *Escherichia coli* KL3, which possesses a mutation in the shikimate dehydrogenase-encoding *aroE* gene (*4*). KL3 is unable to convert DHS into shikimic acid and exports DHS into the growth medium. KL3 has a second copy of the 3-dehydroquinate (DHQ) synthase-encoding *aroB* gene inserted into its 3-phosphoglycerate dehydrogenase-encoding *serA* gene. Insertion of a second copy of *aroB* was needed for compete conversion of DAHP into DHQ (*5*). By inserting *aroB* into *serA*, KL3 was rendered incapable of synthesizing the serine it would require for growth in minimal medium.

KL3 was transformed with pKL4.79B, a plasmid designed to study the impact of DAHP synthase specific activity on DHS synthesis (*4*). As the first committed enzyme of aromatic biosynthesis, DAHP synthase regulates carbon flow into the pathway. Since feedback inhibition of DAHP synthase by the aromatic amino acids is known to be the most significant mode of regulation (*6*), an $aroF^{FBR}$ locus encoding a feedback insensitive isozyme of DAHP synthase (*7*) was inserted into pKL4.79A. Expression of $aroF^{FBR}$ behind the regulatable P_{tac} promoter and inclusion of the gene encoding the regulatory LacI protein ($lacI^q$) on pKL4.79B allowed expression of DAHP synthase to be titrated by addition of IPTG to the culture medium. Maintenance of pKL4.79B in KL3 was ensured by including a functional copy of *serA* in the plasmid. Growth of KL3/pKL4.79B in culture medium lacking serine forces the host strain to maintain the plasmid.

Fed-batch fermentations of KL3/pKL4.79B were performed in a 2-L bench-top fermentor (*4*). Various concentrations of IPTG were added at 6 h intervals. As seen in Table I, DAHP synthase specific activity increased as a function of IPTG addition. Incremental increases in IPTG concentrations led to corresponding improvements in DHS titers (entries 1-4, Table I), reaching a maximum of 52 g/L of DHS (entry 4, Table I) and a 20 % conversion (mol DHS/mol glucose). However, when the IPTG concentration was increased further, a pronounced decrease in DHS titer was observed (entry 5, Table I), implying that there is a level of DAHP synthase activity above which further increases have a negative impact on DHS-synthesizing ability.

Further improvement in DHS titers were realized by addition of a copy of the gene encoding transketolase (*tktA*) to pKL4.79B to form pKL4.124A (*8, 9*). Transketolase is one of two cellular enzymes responsible for formation of D-erythrose 4-phosphate (E4P). An unstable molecule which has been shown to dimerize and polymerize, E4P has never been detected in a living cell (*10*). Fermentation of KL3/pKL4.124A at the IPTG concentration which had previously proven optimal yielded 66.3 g/L of DHS (entry 6, Table I), representing a 28% conversion (mol DHS/mol glucose) (*4*).

Table I. Product titers, yields, and DAHP synthase activities for KL3/pKL4.79B and KL3/pKL4.124A.

Entry	IPTG[c] addition	DAHP spec. act.[d] 24 h	48 h	[DHS] (g/L)	DHS Yield (mol/mol)
1[a]	0.0	.011	.007	24.3	13%
2[a]	0.32	.098	.048	29.1	16%
3[a]	1.6	0.33	0.12	34.0	21%
4[a]	4.8	0.30	0.42	52.0	20%
5[a]	8.0	2.8	1.1	26.8	12%
6[b]	4.8	0.12	.062	66.3	28%

[a]KL3/pKL4.79B
[b]KL3/pKL4.124A
[c]Amount of IPTG (mg) added at 12, 18, 24, 30, 36, and 42 h.
[d]μmol DAHP formed/min mg at 37°C.

Synthesis of DHS from Glucose, Xylose, and Arabinose Mixtures

Starch-derived glucose is the traditional feedstock for microbe-catalyzed synthesis because of it availability and price. However, the metabolic price paid for using the *E. coli* PEP-phosphotransferase system (PTS) of glucose uptake is reflected by the 43% maximum theoretical yield (*4*). In contrast to glucose, D-xylose and L-arabinose uptake by *E. coli* utilize high affinity ATP-driven permeases to phosphorylate glucose (*11*). Because these permeases do not use PEP, the maximum theoretical yield of DHS from either pentose is 71% (mol DHS/mol pentose) (*12*). Abundant supplies of xylose and arabinose are found in corn fiber, which constitutes 10% of the dry weight of the plant (*13*) and is currently disposed of as corn gluten feed, a low value product (*14*). While significant process challenges must be met before high volume streams of xylose and arabinose become available, creation of a high value-added outlet for such streams would provide incentive to overcome these challenges. The 3:3:2 molar ratio of glucose:xylose:arabinose used approximate the composition of pentose streams that could be derived from corn fiber in the future (*15*).

With KL3/pKL4.124A as the biocatalyst, fed-batch fermentations were run using glucose, xylose, and glucose:xylose:arabinose mixtures as carbon sources (*12*). Typical fed-batch fermentations using glucose employ dissolved oxygen sensor-controlled addition of substrate to the fermentation broth (*4*). This addition regimen avoids accumulation of glucose in the culture broth. However, when xylose was used as the carbon source, dissolved oxygen regulated control of substrate addition was not possible. Pentose substrates were therefore under manual control such that accumulation of xylose in the broth was minimized as best as possible. Unfortunately, as the metabolic capacity of the cells began to diminish near the end of these fermentations, accumulation of xylose in the broth was difficult to avoid. To minimize this difficulty, fermentations were run for 42 h rather than 48 h and yield calculations were made after subtraction of residual xylose in the culture medium.

Use of glucose and xylose as independent carbon sources resulted in DHS titers of 46.0 g/L and 43.1 g/L, respectively, reflecting DHS yields of 28% (mol DHS/mol glucose) and 33% (mol DHS/mol xylose). When the 3:3:2 mixture of glucose:xylose:arabinose was used, the titer of DHS synthesized increased to 64.0 g/L while the yield rose to 41% (mol DHS/mol substrate). Taking into account the relative concentrations of each substrate, the maximum theoretical yield for the 3:3:2 mixture is 58% (*12*). In addition to the improved titers and yields, switching to the 3:3:2 mixture of substrates enabled fermentations to be run using dissolved oxygen controlled substrate addition.

Conversion of DHS into Protocatechuic Acid

Although no large-scale industrial synthesis of protocatechuic acid (PCA) is currently practiced, development of an inexpensive process that could be practiced on a large scale might prove useful. The structural backbone of PCA is found in a variety of pharmaceutical products including bronchodilators hexoprenaline and epinephrine and antidepressant rolipram. Processes leading to PCA are also critical to our strategy of using DHS as a chemical building block since PCA provides an entry route to catechol, adipic acid, vanillin, gallic acid, and pyrogallol synthesis (Figure 1).

Microbe-Catalyzed Transformation

Enzyme-catalyzed conversion of DHS to PCA requires DHS dehydratase (Figure 1), an enzyme not native to *E. coli.* The *aroZ* gene encoding DHS dehydratase was isolated from *Klebsiella pneumoniae* (*16*). Creation of a microbial catalyst for high yielding PCA synthesis began with *E. coli* KL7 (*17*). Like KL3, KL7 is unable to convert DHS into shikimic acid. In addition, a cassette encoding *aroB* and *aroZ* was inserted into the *serA* gene of KL7. Inclusion of DHS dehydratase-encoding *aroZ* on the genome of KL7 should channel DHS into PCA synthesis. Plasmid pSK7.54 was prepared which contained $aroF^{FBR}$, *serA* and an additional copy of *aroZ* to supplement DHS dehydratase expression from the genome (*18*). Fed-batch fermentations of KL7/pSK7.54 using glucose as the carbon source resulted in fermentation broths containing 22.5 g/L of PCA after 48 h of culturing, representing a 17% (mol/mol) conversion of glucose into PCA (*18*).

Chemical Transformation

Thermal dehydration of DHS also affords PCA (Figure 1) (*18*). Cell-free fermentation broths containing 62 g/L of DHS were passed through a filter to remove soluble proteins. Refluxing the resulting solution (pH 7) under either an air or nitrogen atmosphere for 2 h afforded a 72-76% (mol/mol) conversion of DHS into PCA, resulting in an overall yield of 23% (mol PCA/mol glucose). Adjusting the pH of the solution to more acidic or more basic pH failed to increase the percent conversion of DHS into PCA.

Comparison of the two strategies for PCA synthesis from DHS depends upon the ultimate goal of the process. If synthesis of PCA as a final product is the goal, the two-step conversion that relies on thermal dehydration of DHS is superior. However, if PCA is needed only as an intermediate which will undergo further catalysis, biocatalytic synthesis using DHS dehydratase is superior, since the need for separate, sequential biocatalytic processes can be avoided.

Conversion of DHS into Vanillin

Vanillin is the second largest food additive used each year. As a consequence of limited natural sources of this ubiquitous flavor and fragrance, the bulk of commercial vanillin is synthesized from phenol (*19*). A two-step synthesis of vanillin from glucose via DHS intermediacy was recently reported (Figure 3) (*17*). DHS dehydratase-catalyzed conversion of DHS into PCA is followed by formation of the methyl ether, catalyzed by catechol-*O*-methyltransferase, an enzyme not native to *E. coli* but for which the gene (*comt*) has been cloned from rat liver (*20*). In a separate step, incubation of the resulting vanillic acid with aryl-aldehyde dehydrogenase (*21*) affords vanillin.

Creation of a biocatalyst for vanillic acid biosynthesis began with KL7, which already expresses DHS dehydratase (*17*). Plasmid pKL5.26A was constructed which contains *aroF*^{FBR}, *serA* and the *comt* gene expressed from P_{tac}. Fed-batch fermentation of KL7/pKL5.26A produced a mixture of products (Table II) (*17*). Formation of vanillic acid and isovanillic acid in a molar ratio of 6:1 reflects the reported preference of the enzyme for *meta*-hydroxyl methylation (*22*) while the appearance of DHS reflects the need for a small increase in DHS dehydratase activity. The dominance of PCA relative to vanillic acid in the product mixture appeared to be an indication of insufficient COMT. However, while inclusion of a second copy of the *comt* gene on the plasmid did raise COMT specific activity two-fold, fermentations using this plasmid produced only a negligible amount of additional vanillic acid. In contrast, when fermentations of KL7/pKL5.26A were supplemented with L-methionine, a precursor of the COMT cofactor *S*-adenosylmethionine, vanillate concentrations nearly doubled to 4.9 g/L at the expense of PCA accumulation (Table II).

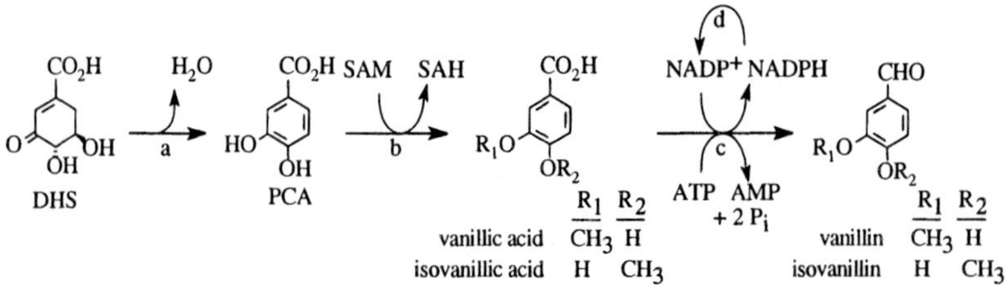

Figure 3. Synthesis of vanillin from DHS. Enzymes: (a) DHS dehydratase (aroZ); (b) catechol-O-methyltransferase (comt); (c) aryl-aldehyde dehydrogenase; (d) glucose 6-phosphate dehydrogenase, glucose 6-phosphate.

Table II. Products synthesized by KL7/pKL5.26A after 48 h.

L-methionine[a]	[vanillic acid] (g/L)	[isovanillic acid] (g/L)	[PCA] (g/L)	[DHS] (g/L)
—	2.5	0.4	9.7	0.9
+	4.9	1.3	7.1	1.0

[a]0.4 g/L added to fermentation every 6 h beginning at 12 h.

SOURCE: Reproduced with permission from Reference 17. Copyright ACS.

Aryl-aldehyde dehydrogenase was purified from *Neurospora crassa* to remove a dehydrogenase that reduces vanillin to vanillyl alcohol (*23*). Vanillic acid, isovanillic acid, and PCA were extracted from a fermentation broth with ethyl acetate. Reprecipitation of the isolated products increased the vanillic acid:PCA ratio from 1:2 to 2.5:1 (mol/mol). Incubation of the resulting mixture with aryl aldehyde dehydrogenase, $NADP^+$, and ATP for 7 h at 30°C resulted in a 92% reduction of vanillic acid to vanillin but only a 33% reduction of PCA to protocatechualdehyde. Vanillin was extracted from the reaction with CH_2Cl_2, affording a product with 10 mol % isovanillin as the only detected contaminant. Based on the concentration of vanillic acid in the fermentation broth, reduction of vanillic acid through purification of vanillin proceeded in 66% overall yield.

Conversion of DHS into Catechol

The market size of catechol is difficult to estimate because a significant portion is produced for a captive market, where it is used directly for synthesis of pharmaceuticals, flavors, agrochemicals, polymerization inhibitors, and antioxidants. Catechol is synthesized industrially from benzene via cumene and phenol intermediacy (*24*). Oxidation of phenol with H_2O_2 affords a mixture of catechol and hydroquinone which are separated by successive distillations.

Microbe-Catalyzed Transformation

Enzyme-catalyzed formation of catechol from DHS requires DHS dehydratase and PCA decarboxylase (Figure 1). Because PCA decarboxylase is not found in *E. coli*, the *aroY* gene encoding this enzyme was cloned from *Klebsiella pneumoniae* (*16*). Creation of a microbial catalyst for synthesis of catechol began with KL7. Plasmid pSK7.126 was constructed which contains *aroF*^{FBR}, *serA*, and *aroY* (*18*). Fed-batch fermentations of KL7/pSK7.126 synthesized 3.0-3.5 g/L of catechol before the cells lost their synthetic capability (*18*). This result likely reflects the toxicity of catechol toward growing cells.

Several strategies might be considered to avoid catechol toxicity. For example, a single organism could be designed in which expression of *aroZ* and *aroY* gene is tightly regulated and delayed until the cells have reached the stationary phase of growth. This strategy, however, requires that the cells are able to reabsorb DHS synthesized early in the fermentation for conversion into catechol later in the fermentation. A second strategy that avoids this requirement would utilize a host strain resistant to high concentrations of catechol. Finally, a method to extract catechol from the medium before it exceeds 3 g/L could be developed, although this approach would likely increase the expense of the process significantly.

Combined Chemical and Microbe-Catalyzed Transformation

A multistep approach for catechol synthesis has also been developed that uses both chemical and microbial catalysis to avoid toxicity of PCA and catechol. This approach relies upon the high-yielding biocatalysts that have already been developed for high titer DHS synthesis combined with thermal dehydration conditions developed for conversion of DHS into PCA. Design of a biocatalyst specific for PCA decarboxylation would then result in catechol formation. RB791 *serA::aroB*/pSK6.234, was prepared which contained *serA* and *aroY* on the plasmid (*18*). The strain was cultured in a fermentor until the cells had reached the stationary phase of growth (24 h). The fermentation vessel was then sparged with nitrogen and glucose addition was terminated before PCA was added to a final concentration of 17 g/L. After 14 h, all of the PCA was consumed, affording catechol in an 82% (mol/mol) conversion (*18*).

Conversion of DHS into Adipic Acid

Adipic acid is a six-carbon diacid used primarily in production of nylon-6,6. The annual global demand for adipic acid is in the range of 2 billion kilograms, and it is perennially among the top 50 chemicals produced in the U.S. each year. Adipic acid is manufactured from benzene, where the final step of the reaction sequence requires nitric acid oxidation of cyclohexanone and cyclohexanol. Nitrous oxide, a byproduct of the oxidation, has been implicated in ozone layer depletion and the greenhouse effect. Adipic acid production has been estimated to account for approximately 10% of the annual increase in atmospheric nitrous oxide levels (*25*).

Synthesis of adipic acid from DHS utilizes both PCA and catechol as biosynthetic intermediates and requires additional expression of catechol 1,2-dioxgenase in the biocatalytic organism (Figure 1). Since catechol 1,2-dioxygenase is not found in *E. coli*, the *catA* gene encoding the enzyme has been obtained from *Acinetobacter calcoaceticus*. Catechol 1,2-dioxygenase catalyzes conversion of catechol to *cis, cis*-muconic acid, which is readily reduced in a separate step to afford adipic acid. Perhaps the most important characteristic of a successful catalyst for adipic acid synthesis will be the avoidance of catechol accumulation.

Creation of a biocatalytic organism that is able to be grown in a fermentation vessel is still under development. However, biocatalyst AB2834/pKD136/pKD8.243A/pKD8.292 has already been tested in shake flask experiments (*26*). Although cell densities and product titers achieved in shake flasks pale in comparison to those achieved under the more optimal conditions found in a fermentor, validity of the approach can still be established. Genes encoding transketolase (*tktA*), DAHP synthase (*aroF*), DHQ synthase (*aroB*), DHS dehydratase (*aroZ*), PCA decarboxylase (*aroY*), and catechol 1,2-dioxygenase (*catA*) were each inserted into one of the three plasmids while host AB2834 possessed a mutation in the *aroE* locus. Analysis of the culture supernatant of AB2834/pKD136/pKD8.243A/pKD8.292 revealed that on a 1-L scale in laboratory shake flasks, 56 mM glucose was converted into 16.8 mM *cis, cis*-muconic acid (*26*). PCA and catechol were not detected in the culture supernatant. Addition of 10% platinum on carbon to the cell-free culture supernatant followed by hydrogenation at 50 psi for 3 h afforded a 90% conversion of *cis, cis*-muconic acid to adipic acid (*26*).

Conversion of DHS into Gallic Acid

Gallic acid is an important starting material used in the synthesis of a wide range of food additives and pharmaceuticals. Gallic acid is currently obtained from natural sources, including gall nuts harvested in China and the seed pod of a tree found in Peru (*27*). Hydrolysis of the gallotannins prevalent in these sources affords gallic acid.

Microbe-Catalyzed Transformation

Enzyme-catalyzed formation of gallic acid from DHS will rely upon PCA intermediacy (Figure 1). While no naturally occurring enzyme has been reported for direct hydroxylation of PCA, a mutant isozyme of the *Pseudomonas auruginosa* enzyme *p*-hydroxybenzoate hydroxylase has been reported which hydroxylates PCA to afford gallic acid (*28*). Expression of this mutant enzyme encoded by the *pobA** locus in an organism designed for PCA biosynthesis was expected to result in gallic acid formation.

Creation of a microbial catalyst for synthesis of gallic acid began with KL7. Plasmid pSK6.161 was constructed which contained *aroF*^{FBR}, *serA*, P_{tac}*pobA**, and *lacI*^{q}. Expression of the *pobA** locus was regulated through addition of IPTG to the fermentation medium. Under glucose-rich fermentation conditions, KL7/pSK6.161 synthesized 20 g/L of gallic (*29*). While significant improvement in KL7/pSK6.161 is needed to increase the process from its current 12% overall yield toward the maximum theoretical yield of 43%, the basic route for microbial gallic acid synthesis is in hand.

Chemical Transformation

To take full advantage of the aforementioned high-yielding, high-titer microbial syntheses of DHS, the chemical oxidation of DHS to gallic acid has been examined. DHS is an example of an α-hydroxycarbonyl. Historically, this functionality has been oxidized to an α-diketone using Cu^{+2}. Haslam has reported that Cu^{+2}-containing Fehling solution oxidizes DHS to gallic acid (*30*) although we have found that this reaction leads to low yields of gallic acid concurrent with formation of high levels of protocatechuic acid contamination. Reexamination of the reaction of Cu^{+2} with DHS has revealed that high yields of gallic acid can be achieved when acetic acid is used as the reaction solvent (*31*). Stoichiometric quantities of $Cu(OAc)_2$ oxidize DHS in acetic acid solution to gallic acid in 74% yield. Catalytic quantities of $Cu(OAc)_2$ can also be employed for such oxidations when O_2 or air are recruited as a cooxidant. Oxidations of DHS in acetic acid using $Cu(OAc)_2$ lead to formation of only small amounts of protocatechuic acid contamination.

Conversion of DHS into Pyrogallol

Global production of pyrogallol stands at approximately 2 x 10^5 kilograms per year and is produced via thermal decarboxylation of gallic acid in copper autoclaves (*27*). Among its many uses, pyrogallol is used as a photographic developing agent as well as a synthetic starting material, providing the structural backbone of molecules including the muscle relaxant gallamine triethiodide and the insecticide bendiocarb.

Enzyme-catalyzed synthesis of pyrogallol from gallic acid requires gallate decarboxylase. While gallate decarboxylase from *Pantoea agglomerans* T71 is well-characterized (*32*), PCA decarboxylase from *K. pneumoniae* was also found to act on gallic acid to afford pyrogallol (*29*). Attempts to create a single catalytic organism for pyrogallol synthesis by expressing *aroZ*, *pobA**, and *aroY* proved unsuccessful in fed-batch fermentation (*18*). Accumulation of catechol indicated that PCA decarboxylase catalyzed the formation of catechol better than *p*-hydroxybenzoate hydroxylase catalyzed the formation of gallic acid.

As an alternative to using a single biocatalytic organism, two separate biocatalytic steps were utilized for pyrogallol synthesis (*29*). RB791*serA::aroB*/pSK6.234 expressing PCA decarboxylase was cultured in a fermentor until the cells reached the stationary phase of growth (24 h). The fermentation vessel was then sparged with nitrogen and glucose addition was terminated before gallic acid-containing fermentation broth produced by KL7/pSK6.161 was added to the vessel. After 14 h, a 93% conversion of gallic acid to pyrogallol was achieved (*29*).

Concluding Remarks

The biocatalytic conversion of glucose into DHS and subsequent biocatalytic and chemical catalytic conversion of DHS into PCA, vanillin, catechol, adipic acid, gallic acid, and pyrogallol establishes an example of a rudimentary manufacturing tree based on renewable feedstocks. Hydroaromatic DHS is the key intermediate building block derived from glucose, xylose, and arabinose. These carbohydrates, in turn, are derived from starch, hemicellulose or cellulose. The products of this synthetic tree include examples of ultrafine (gallic acid, pyrogallol), fine (protocatechuic acid, vanillin), pseudocommodity (catechol), and commodity (adipic acid) chemicals.

Given ongoing genome sequencing and functional genomic efforts, a vast number of biosynthetic connections will likely be made between carbohydrate starting materials and organic chemicals of commercial significance. Chemists need not be limited by extant biosynthetic pathways. By combining enzyme activities isolated from different organisms, new biosynthetic pathways can be created that have no parallel in nature. Naturally-occurring and unnatural biosynthetic pathways can then be interfaced with catalytic and stoichiometric chemical conversions. The net result may well be that most of the vast array of chemicals currently synthesized from petroleum feedstocks can be synthesized from renewable feedstocks in a cost-effective fashion.

References

1. Szmant, H. H. *Organic Building Blocks of the Chemical Industry*; Wiley & Sons: New York, NY, 1989; p 4.
2. Pittard, A. J. In *Escherichia coli and Salmonella: Cellular and Molecular Biology*; 2nd Ed., Neidhardt, F. C., Ed.; ASM Press: Washington, DC; Vol. 1, pp 458-484.
3. Richman, J. E.; Chang, Y.-C.; Kambourakis, S.; Draths, K. M.; Almy, E.; Snell, K. D.; Strasburg, G. M.; Frost, J. W. *J. Am. Chem. Soc.* **1996**, *118*, 11587-11591.
4. Li, K.; Mikola, M. R.; Draths, K. M.; Worden, R. M.; Frost, J. W. *Biotechnol. Bioeng.* **1999**, *64*, 61-73.
5. Snell, K. D.; Draths, K. M.; Frost, J. W. *J. Am. Chem. Soc.* **1996**, *118*, 5605-5614.
6. Ogino, T.; Garner, C.; Markley, J. L.; Herrmann, K. M. *Proc. Natl. Acad. Sci., USA* **1982**, *79*, 5828-5832.
7. Mikola, M. R.; Widman, M. T.; Worden, R. M. *Appl. Biochem. Biotechnol.* **1998**, *70*, 905-918.
8. Draths, K. M.; Pompliano, D. L.; Conley, D. L.; Frost, J. W.; Berry, A.; Disbrow, G. L.; Staversky, R. J.; Lievense, J. C. *J. Am. Chem. Soc.* **1992**, *114*, 3956-3962.
9. Patnaik, R.; Liao, J. C. *Appl. Environ. Microbiol.* **1994**, *60*, 3903-3908.
10. Williams, J. F.; Blackmore, P. F.; Duke, C. C.; MacLeod, J. K. *Int. J. Biochem.* **1980**, *12*, 339-344.
11. Lin, E. C. C. In *Escherichia coli and Salmonella: Cellular and Molecular Biology*; 2nd Ed., Neidhardt, F. C., Ed.; ASM Press: Washington, DC; Vol. 1, pp 307-342.
12. Li, K.; Frost, J. W. *Biotechnol. Prog.* **1999**, *15*, 876-883.
13. Moniruzzaman, M.; Dien, B. S.; Ferrer, B.; Hespell, R. B.; Dale, B. E.; Ingram, L. O. Bothast, R. J. *Biotechnol. Lett.* **1996**, *18*, 985-990.
14. May, J. B. In *Corn: Chemistry and Technology*; Watson, S. A., Ramstad, P. E., Eds.; American Association of Cereal Chemists: St. Paul, MN, 1987; pp 377-397.
15. Carlson, T. *Preliminary Program Proceedings*, Corn Utilization Conference V, June 8-10, 1994; National Corn Growers Association: St. Louis, MO.
16. Draths, K. M.; Frost, J. W. *J. Am. Chem. Soc.* **1995**, *117*, 2395-2400.
17. Li, K.; Frost, J. W. *J. Am. Chem. Soc.* **1998**, *120*, 10545-10546.
18. Kambourakis, S. Ph.D. Thesis, Michigan State University, 2000.

19. Esposito, L.; Formanek, K.; Kientz, G.; Mauger, F.; Maureaux, V.; Robert, G.; Truchet, F. In *Kirk-Othmer Encyclopedia of Chemical Technology*, 4th Ed.; Kroschwitz, J. I., Howe-Grant, M., Eds.; Wiley & Sons: New York, NY, 1997; Vol. 24, p 812.
20. Salminen, M.; Lundstrom, K.; Tilgmann, C.; Savolainen, R.; Kalkkinen, N.; Ulmanen, I. *Gene* **1990**, *93*, 241-247.
21. Gross, G. G.; Bolkart, K. H.; Zenk, M. H. *Biochem. Biophys. Res. Commun.* **1968**, *32*, 173-178.
22. Guldberg, H. C.; Marsden, C. A. *Pharmacol. Rev.* **1975**, *27*, 135.
23. Gross, G. G.; Zenk, M. H. *European. J. Biochem.* **1969**, *8*, 413-419.
24. Szmant, H. H. *Organic Building Blocks of the Chemical Industry*; Wiley & Sons: New York, NY, 1989; pp 512-519.
25. Thiemens, M. H.; Trogler, W. C. *Science* **1991**, *251*, 932.
26. Draths, K. M.; Frost, J. W. *J. Am. Chem. Soc.* **1994**, *116*, 399-400.
27. Leston, G. In *Kirk-Othmer Encyclopedia of Chemical Technology*, 4th Ed.; Kroschwitz, J. I., Howe-Grant, M., Eds.; Wiley & Sons: New York, NY, 1996; Vol. 19, p 778.
28. Entsch, B.; Palfey, B. A.; Ballou, D. P.; Massey, V. *J. Biol. Chem.* **1991**, *266*, 17341.
29. Kambourakis, S.; Draths, K. M.; Frost, J. W., submitted for publication.
30. Haslam, E.; Haworth, R. D.; Knowles, P. F. *J. Chem. Soc.* **1961**, 1854.
31. Kambourakis, S.; Frost, J. W., submitted for publication.
32. Zeida, M.; Wieser, M.; Yoshida, T.; Sugio, T.; Nagasawa, T. *Appl. Environ. Microbiol.* **1998**, *64*, 4743.

Chapter 12

Synthesis and Properties of Polymers Derived from Substituted Lactic Acids

Mao Yin, Tara L. Simmons, and Gregory L. Baker*

Department of Chemistry, Crop and Food Bioprocessing Center, and Center for Fundamental Materials Research, Michigan State University, East Lansing, MI 48824

The preparation of substituted glycolides can greatly expand the pool of monomers available for the synthesis of polyglycolides, and provide a broader range of physical properties for degradable polymers. Ethyl-, hexyl-, isobutyl-, and benzylglycolide were prepared and converted to high molecular weight polymers by ring opening polymerization in solution and using melt techniques. The polymers were clear amorphous materials with glass transition temperatures ranging from 55 to –22 °C. For alkyl-substituted poly(glycolide)s, increasing the length of a linear alkyl chain decreased T_g, while introducing branches in the alkyl chain increased T_g. Above 320 °C, the polymers thermally degrade to monomer by depolymerization. Preliminary degradation studies at 55 °C in H_2O at pH 7.4 show that substituted poly(glycolide)s degrade slower than polylactide, with poly(isobutylglycolide) the most stable polymer toward hydrolysis.

Background

Polymers derived from "green" sources are playing an increasingly important role in the U.S. polymer industry. Green polymers have the advantage of being derived from renewable sources, and in most cases, degrade to benign products at the end of their useful life. Two classes of polymers have received the most attention, the poly(β-hydroxybutyrate)s (PHBs) and polylactic acid (polylactide). (*1*) Both are obtained via fermentation processes. In the case of PHB, a bacterial fermentation converts alkanoic acids to high molecular weight β-hydroxybutyrate polymers which are stored in the cells as an energy reserve. For polylactic acid, fermentation of starch yields L-lactic acid, which after conversion to the cyclic dimer, is polymerized via ring opening polymerization to high molecular weight polylactide. One of the most important properties of polylactides is that they degrade by hydrolysis and are bioresorbable, making them important materials for applications in medicine as sutures and surgical implants. (*2,3*)

A recent study analyzed the energy inputs associated with the production of polymers from β-hydroxyacids. (*4*) The initial findings suggest that for the PHBs, the principal advantage of these green materials lies in their degradability; the energy inputs into their synthesis from nonrenewable sources appear to at least match those avoided by using natural materials. Because of their cost, PHBs have had limited commercial success to date. Polylactides apparently have more favorable economics, and with the commercialization of polylactide by Cargill Dow (*5*) on the 100 million lb./year scale, polylactides are poised to compete in large volume markets that are now the domain of thermoplastic polymers derived from petrochemical sources.

Two α-hydroxyacids currently make up the polylactide family: lactic acid and glycolic acid. As illustrated in Figure 1 for glycolide, both are polymerized by ring opening polymerization using catalysts like Sn(2-ethylhexanoate)$_2$ and Al(OiPr)$_3$. The physical properties and degradation rates of polylactides are typically controlled by copolymerization with monomers such as glycolide (*6,7*) and ε-caprolactone, (*8*) or by formation of block copolymers. (*9,10,11,12,13,14,15*) Alternatively, the degree of crystallinity in polylactides can be manipulated to achieve similar results. (*16,17*) Lactide is available as 3 diastereomers, the R,R, S,S, and the R,S (or *meso*) isomers. When prepared from a single enantiomer, polylactides can be highly crystalline with melt temperatures near 180 °C. (*18,19*) Polymer crystallinity declines as the stereoregularity of the polymer chain decreases, and thus polymerizations of the meso lactide or a 1:1 mixture of the R,R and S,S isomers (*rac*-lactide) provide amorphous materials with T_gs around 50 °C. (*18*) However, recent developments in chiral polymerization catalysts now enable preferential polymerization of the L-lactide from rac-lactide, (*20*) the formation of

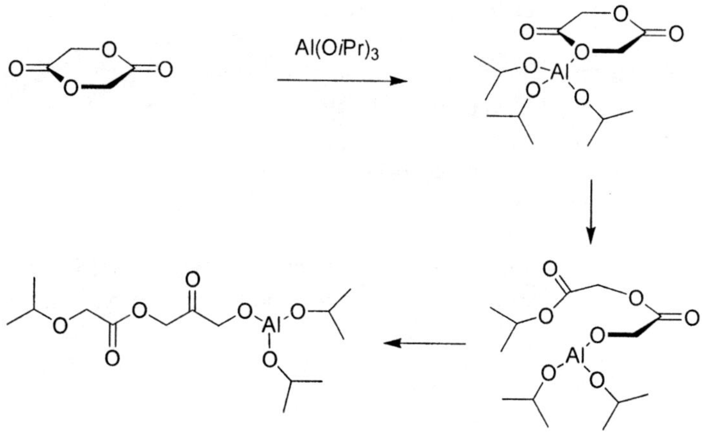

Figure 1. Ring opening polymerization of glycolides. $Al(OiPr)_3$ initiates the growth of three polymer chains. Only one is shown for clarity.

stereoregular polymers from the meso lactide, (*21*) and the direct formation of a high melting stereocomplex of poly(D-lactide) and poly(L-lactide) from *rac*-lactide. (*22*)

In terms of physical properties, having only a methyl group pendant to the polymer backbone is limiting, and to obtain a broader range of physical properties for polyglycolides we turned to a simple analogy based on polyolefins. Relatively minor changes in the substituents attached to the polymethylene chain drastically alter the physical properties of polyolefins, and thus analogous changes to the polyglycolide backbone should provide degradable polymers with a broad range of physical properties. Ideally, we would like to be able to devise degradable analogs of commercial thermoplastics and thermosets from renewable resources. In this overview, we describe our initial efforts in this area with a focus on the synthesis and characterization of substituted polyglycolides.

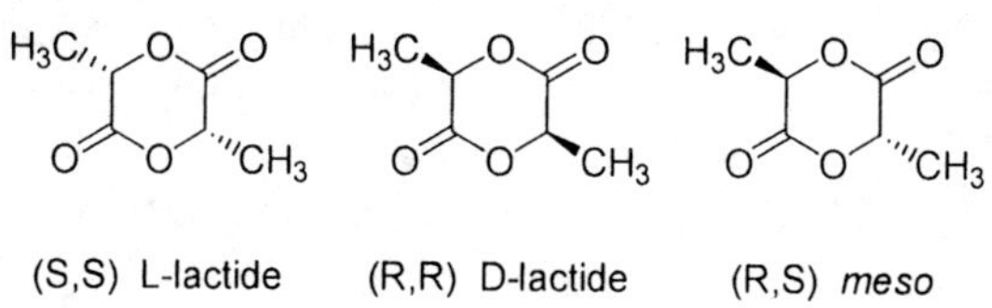

Figure 2. The three lactide diastereomers.

Raw Materials for Substituted Polyglycolides

For our approach to be useful, we need a "green" source for the monomers. One such source is the biosynthetic pathway for the synthesis of α-amino acids. As an example, consider the biosynthetic pathway to L-phenylalanine shown in Figure 3. The last step in the synthesis is the amination of the α-ketoacid, but interception and reduction of the ketone would provide phenyllactic acid, a substituted lactic acid that could be converted to its cyclic dimer and polymerized. Frost has shown that the biosynthetic pathway can be altered to favor production of the desired α-hydroxyacid, and proper choice of the enzyme

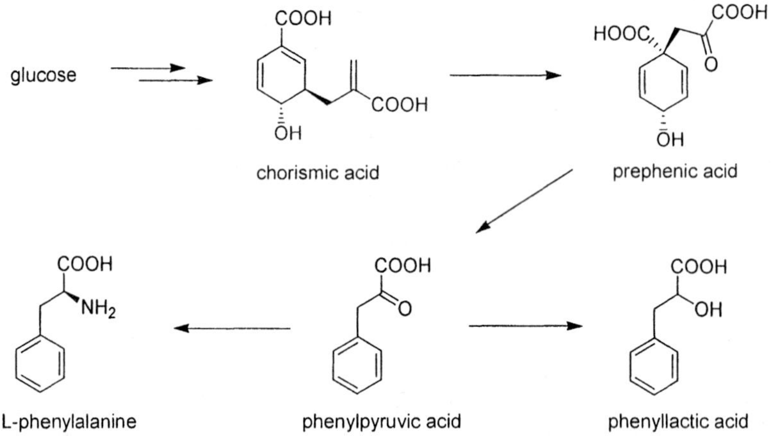

Figure 3. Biosynthetic pathway to phenylalanine and phenyllactic acid.

for reduction of phenylpyruvic acid yields either the D or L enantiomer. (*23*) Polymers derived from single enantiomers of phenyllactic acid should provide crystalline polymers. Following this approach, α-hydroxyacids can be obtained that have the same carbon skeletons as the naturally occurring amino acids. For example, the pathways to valine and leucine could, in principle, form the basis of a fermentation process that ultimately yields polyglycolides with branched side chains (Figure 4).

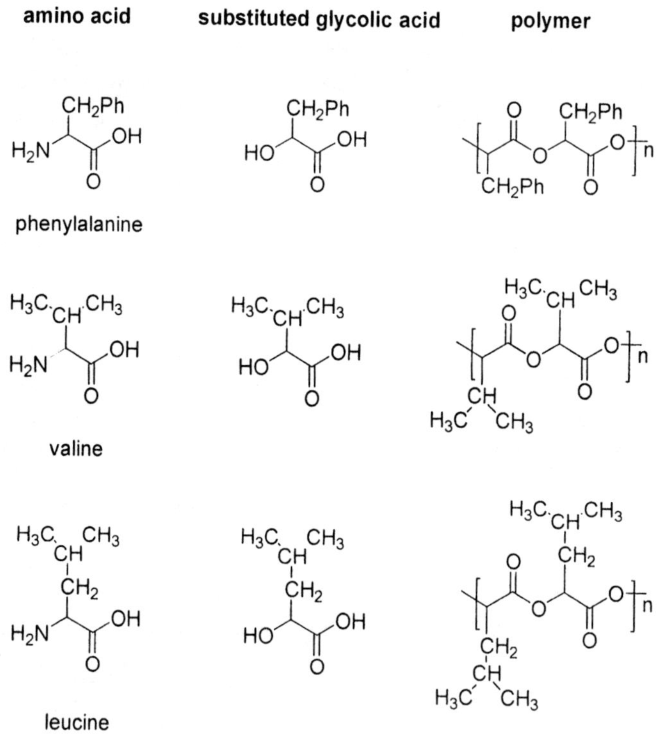

Figure 4. Examples of glycolic acids and polyglycolides that can be obtained from amino acid biosynthetic pathways.

In addition to intercepting and reducing α-ketoacid intermediates formed in amino acid syntheses, α-ketoesters prepared by traditional synthetic methods can be enzymatically reduced to give optically active α-hydroxyacids. Using Baker's yeast, α-ketoesters were reduced in ees as high as 90%. (*24*) A complementary approach is the enzymatic oxidation of alkanoic acids. Several examples reported to date have high chemical yields (99%) and high enantioselectivity (99 % ee). (*25*)

Our initial focus is on four simple alkyl derivatives of lactic acid, with an emphasis on discovering trends in the T_gs and degradation rates of the amorphous materials. Preliminary data on a fifth polymer, poly(benzylglycolide), are included for comparison. The structures of the monomers are shown in Figure 5. The needed α-hydroxyacids were obtained from commercial sources, and were converted to the corresponding cyclic dimers by the standard procedure used to prepare lactide. (*26*) Acid-catalyzed

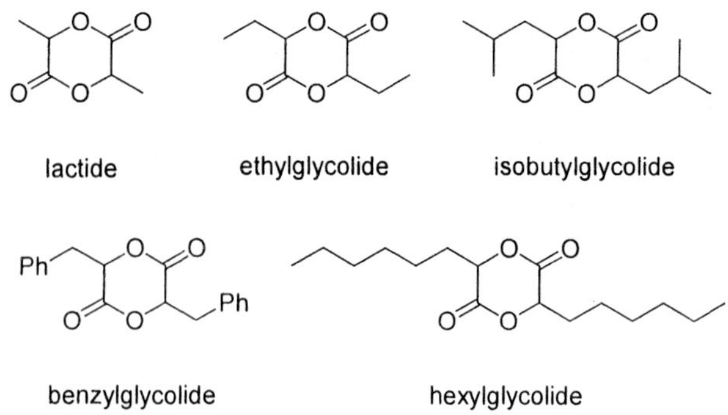

Figure 5. The five glycolide derivatives used in this study.

esterification of the acids gave a mixture of dimers and linear polymers. The dimers were isolated and the low molecular weight polymer was heated under vacuum in the presence of a transesterification catalyst, allowing collection of the volatile dimer as it was formed. The yields in these reactions were good (60-70 %), but similar reactions optimized for lactic acid give 95 % conversions to the dimer. Because the syntheses used racemic α-hydroxyacids, the glycolides obtained were mixtures of the R,R; S,S and R,S diastereomers.

Melt Polymerizations

The dimers were converted into high molecular weight linear polymers by ring opening polymerization under melt conditions. Melt polymerizations, have several advantages over solution polymerizations and are widely used for the synthesis of lactides. (*27*) The high monomer concentration in a melt favors rapid propagation, and the lack of solvent minimizes environmental costs associated with recovery of the solvent. The catalyst system used was Sn(2-ethylhexanoate)$_2$ with *t*-butylbenzyl alcohol as the initiator. Experimental details of the polymerizations are provided elsewhere. (*26*)

Shown in Figure 6 are data for the melt polymerization of four alkyl derivatives of glycolide at 130 °C and a monomer/initiator ratio of 100. M_n increases linearly with conversion, as would be expected for a living polymerization, and the slopes of the lines are proportional to the molecular weight of the monomer. At long polymerization times the molecular weight begins to decrease, presumably as a result of intramolecular transesterification (back biting) reactions that form cyclic oligomers. (*28*) Polymerizations typically reach a maximum conversion of ≈97%, a consequence of the equilibrium nature of the ring opening polymerization of lactide. (*27*) The molecular weights of the polymers depend on the monomer/initiator ratio, and melt polymerizations of the alkyl glycolides can reach M_n > 100,000 g/mol.

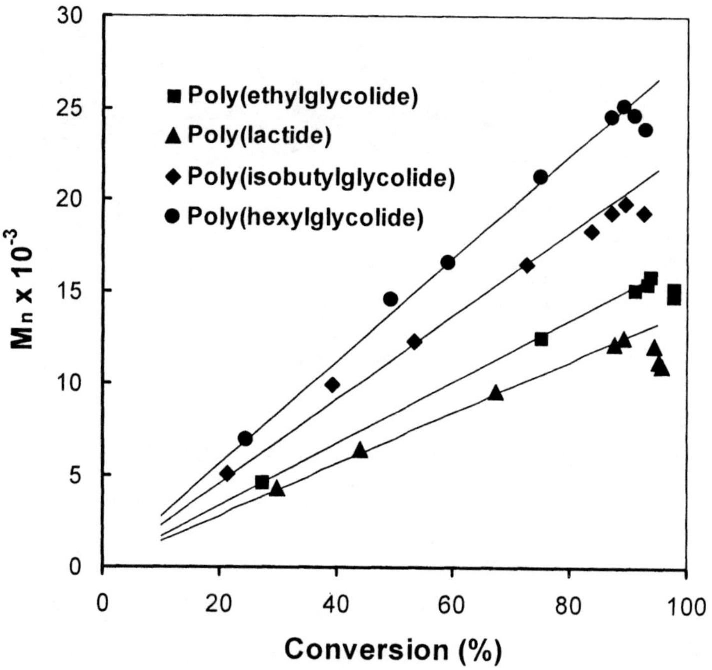

Figure 6. Melt polymerization of alkyl-substituted glycolides at 130 °C, adapted from reference 26.

Figure 7 shows that the kinetic data for these polymerizations follow the expected trend; as the steric bulk of the ring substituent increases, the rate of propagation decreases. Thus, isobutylglycolide has the slowest polymerization rate and lactide has the fastest. The lines through the data points are based on a model that assumes equilibrium control of the polymerization, with an equilibrium monomer concentration of 3%. The calculated line and experimental points match well, allowing extraction of the relative rate constants for propagation from the data. The relative rates are lactide: 2.2, ethylglycolide: 1.6, hexylglycolide: 1.2, and isobutylglycolide: 1. The rate of ring opening polymerization for benzylglycolide is significantly slower than that of lactide. A quantitative comparison with the other glycolides is not yet possible since benzylglycolide polymerizations are run at 180 °C due to the higher melting point of the monomer.

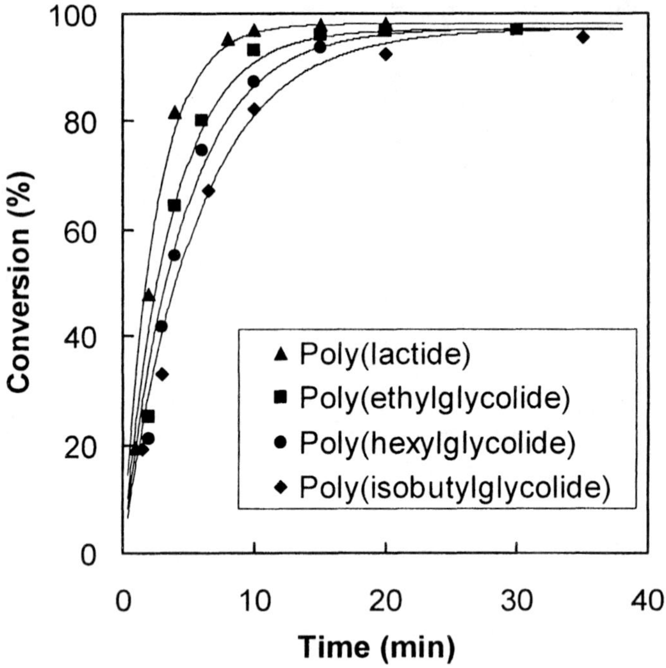

Figure 7. Kinetic data for the melt polymerization of alkyl-substituted glycolides. Adapted from reference 26.

Solution Polymerizations

Because of lower temperatures and better mixing, solution polymerizations tend to be more controlled than melt reactions. For example, $Al(O\textit{i}Pr)_3$ initiated solution polymerizations give predictable molecular weights with narrow molecular weight distributions. (*28*) For the alkylglycolide polymerizations we have studied to date, most show a linear relationship between molecular weight and conversion. Another consequence of a solution polymerization is that the rate of propagation is greatly reduced due to the lower monomer concentration and the lower temperatures typically used in solution polymerizations. With increased substitution on the glycolide ring, the polymerizations of substituted glycolides can take days instead of the minutes typical for melt polymerizations. An example of the solution polymerization of a sterically hindered glycolide, benzylglycolide, is shown in Figure 8. Two characteristics of benzylglycolide

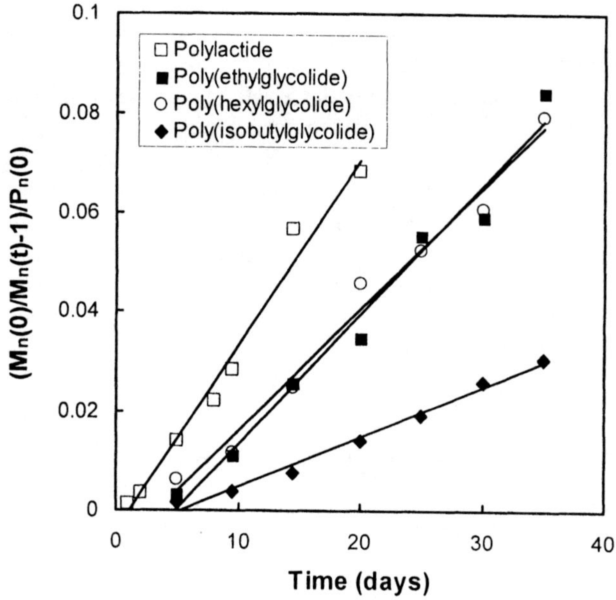

Figure 8. Degradation of alkyl-substituted polyglycolide at 55 °C in water, at pH 7.4. The data are plotted in terms of 1/(degree of polymerization) to give the degradation rate as the slope of the data.

polymerizations are immediately obvious. The reaction times are measured in hundreds of hours, and the conversions appear to approach limiting values. The lower monomer concentrations of benzylglycolide solution polymerizations (0.1 M vs. 7.8 M for a melt polymerization) have the major effect on the rate of polymerization; the lower temperatures (100 °C vs. 130 °C for melt polymerizations) also play a role. These polymerizations also show characteristics of equilibrium polymerization, the effects of which are more pronounced in solution polymerizations due to the low monomer concentrations. The data of Figure 8 imply an equilibrium monomer concentration of 0.07 M for benzylglycolide polymerizations at 70 °C. This compares with 0.055 M reported for the polymerization of lactide at 80 °C.

Polymer Properties

The properties of representative polymers obtained from melt polymerizations are shown in Table 1. These data are from polymerizations run at a monomer/initiator ratio of 200. The higher monomer/initiator ratio produces higher molecular weight materials, which are better suited for measuring physical properties. The polydispersities for the polymers, obtained from GPC measurements in THF using polystyrene standards to calibrate the molecular weights, are higher than expected for a polymerization with living character. This result follows from the intramolecular transesterification reactions that effectively compete with propagation at high conversions (e.g. see Figure 6). The polymers are soluble in a broad range of solvents including THF, chloroform, methylene chloride, toluene, benzene, acetone, and ethyl acetate. Clear colorless films can be cast from each polymer, although films of poly(ethylglycolide) and poly(hexylglycolide), with T_gs below room temperature, tended to be tacky.

Table 1. Properties of Alkyl-Substituted Polyglycolides

Polymer	M_n	M_n/M_w	T_g (°C)	T_d(°C)
Poly(lactide)	35,200	1.89	66	324
Poly(ethylglycolide)	45,600	1.78	15	344
Poly(hexylglycolide)	43,200	1.91	-37	362
Poly(isobutylglycolide)	47,300	1.83	22	362
Poly(benzylglycolide)[a]	57,500	1.64	50	359

a. run at a monomer/initiator ratio of 500

All of the polymers obtained from alkylglycolides were amorphous, as expected given the mixture of diastereomers used in the polymerizations. Surprisingly, poly(benzylglycolide) also was amorphous even though a single enantiomer, L-phenyllactic acid, was used to prepare the glycolide. Inspection of the ^1H NMR showed that 5-15 % of the stereocenters racemized during the polymerization, and this stereo-irregularity likely frustrates crystallization. The polymer T_gs were measured under He at 10 °C/min using a Perkin-Elmer DSC-7. As shown in Table 1, the T_gs decreased as the length of the pendant side chain increased. Linear alkyl side chains present nearly the same rotational barrier for changes in the backbone conformation, and thus changes in the length of the pendant chain should cause only minor perturbations to T_g. However, lengthening the pendent chain increasingly screens the polar polyester chains from each other, effectively minimizing chain-chain dipolar effects. Thus, the T_gs should decrease in the order poly(lactide) > poly(ethylglycolide) > poly(hexylglycolide). Branched side chains or those that contain sterically demanding groups such as aryl rings increase the backbone rotational barriers, stiffen the polymer chains, and cause an increase in T_g. The data for poly(isobutylglycolide) and poly(benzylglycolide) follow that trend.

Apparently, the polymers all degrade by depolymerization. During TGA runs (40 °C/min, N_2 atmosphere), the onset for weight loss was followed by complete and near linear weight loss over a 40° range. For each sample, the onset for decomposition, T_d, was defined by the intersection of the tangent to the weight loss curve with 100 weight %. The T_d values were sample dependent, presumably due to catalyst residues that enhance the depolymerization rate. Samples not purified by repeated precipitation into methanol decomposed at lower temperatures, but purified samples gave constant degradation profiles. Supporting depolymerization were controlled thermal decompositions of poly(benzylglycolide); monomer was recovered in >90% yield. The T_d values reflect the volatility of the monomer. Thus, poly(lactide) has the lowest T_d value and poly(isobutylglycolide) and poly(hexylglycolide) the highest.

Degradation experiments were run at 55 °C at pH 7.4. As shown in Figure 8 polylactide degrades the fastest, and poly(isobutylglycolide) degrades the slowest. Since the experiments were run above the T_g of each material, one would expect the rate of degradation to be related to the hydrophobicity of the polymers. We calculated the solubility parameters for each polymer using the group contribution method. Poly(isobutylglycolide) is the most hydrophobic ($\delta = 7.8$) and lactide is the most hydrophilic ($\delta = 8.6$). However, the value for ethylglycolide ($\delta = 8.5$) is nearly the same as lactide, and thus a simple argument based on solubility parameters alone is unlikely to satisfactorily explain the trends in the data. Preliminary data for poly(benzylglycolide) show that it degrades even slower than poly(isobutylglycolide). One conclusion from the degradation study is that adding alkyl-substituted glycolides to lactide polymerizations would be an effective strategy for decreasing the degradation rate of polylactides.

Conclusions

A variety of substituted glycolic acids can be derived from natural sources, and our work with several such derivatives indicates that they can readily be converted to cyclic dimers and polymerized to high molecular weight polymers using the technology that currently exists for polylactide. Converting the methyl group of lactide to a larger substituent results in slower polymerization rates and slower polymer degradation rates. These polymers have different physical properties than polylactide, and proper choice of the starting glycolic acid should lead to degradable polymers with physical properties that mimic those of polymers derived from petroleum. The polymers prepared to date were amorphous, but the use of monomers derived from a single enantiomer should lead to crystalline polymers with a broader range of properties.

References

1. Chiellini, E.; Solaro, R. *Advanced Materials* **1996**, *8*, 305-313.
2. Vert, M.; Li, S. M.; Spenlehauer, G.; Guerin, P. *Journal of Materials Science-Materials in Medicine* **1992**, *3*, 432-446.
3. Kricheldorf, H. R.; Kreiser-Saunders, I. *Macromolecular Symposia* **1996**, *103*, 85-102.
4. Gerngross, T. U. *Nature Biotechnology* **1999**, *17*, 541-544.
5. Thayer, A. *Chem. Eng. News* **1997**, *75*, 14-16.
6. Grijpma, D. W.; Nijenhuis, A. J.; Pennings, A. J. *Polymer* **1990**, *31*, 2201-2206.
7. Joziasse, C. A. P.; Veenstra, H.; Topp, M. D. C.; Grijpma, D. W.; Pennings, A. J. *Polymer* **1998**, *39*, 467-474.
8. Grijpma, D. W.; Zondervan, G. J.; Pennings, A. J. *Polymer Bulletin* **1991**, *25*, 327-333.
9. Deng, X. M.; Xiong, C. D.; Cheng, L. M.; Xu, R. P. *Journal of Polymer Science Part C-Polymer Letters* **1990**, *28*, 411-416.
10. Bachari, A.; Belorgey, G.; Helary, G.; Sauvet, G. *Macromolecular Chemistry and Physics* **1995**, *196*, 411-428.
11. Chin, I. J.; Kim, K. S.; Chung, S. N.; Lee, S. Y.; Kim, M. N.; Yoon, J. S. *Abstracts of Papers of the American Chemical Society* **1998**, *216*, U33-U33.
12. Deng, X. M.; Xiong, C. D.; Cheng, L. M.; Huang, H. H.; Xu, R. P. *Journal of Applied Polymer Science* **1995**, *55*, 1193-1196.
13. Deng, X. M.; Zhu, Z. X.; Xiong, C. D.; Zhang, L. L. *Journal of Polymer Science Part a-Polymer Chemistry* **1997**, *35*, 703-708.
14. Li, S. M.; Rashkov, I.; Espartero, J. L.; Manolova, N.; Vert, M. *Macromolecules* **1996**, *29*, 57-62.
15. Song, C. X.; Feng, X. D. *Macromolecules* **1984**, *17*, 2764-2767.
16. Wang, J. L.; Kean, R.; Randall, J.; Giles, D. *International Journal of Polymer Analysis and Characterization* **1998**, *4*, 393-405.
17. Perego, G.; Cella, G. D.; Bastioli, C. *Journal of Applied Polymer Science* **1996**, *59*, 37-43.
18. Starkweather, H. W.; Avakian, P.; Fontanella, J. J.; Wintersgill, M. C. *Macromolecules* **1993**, *26*, 5084-5087.
19. Miyata, T.; Masuko, T. *Polymer* **1998**, *39*, 5515-5521.
20. Wisniewski, M.; LeBorgne, A.; Spassky, N. *Macromolecular Chemistry and Physics* **1997**, *198*, 1227-1238.
21. Ovitt, T. M.; Coates, G. W. *J. Am. Chem. Soc.* **1999**, *121*, 4072-4073.
22. Radano, C. P.; Baker, G. L.; Smith, M. R. *J. Am. Chem. Soc.* **2000**, *122*, 1552-1553.
23. Snell, K. D.; Draths, K. M.; Frost, J. W. *J. Am. Chem. Soc.* **1996**, *118*, 5605-5614.

24. Kayser, M. M.; Mihovilovic, M. D.; Kearns, J.; Feicht, A.; Stewart, J. D. *Journal of Organic Chemistry* **1999**, *64*, 6603-6608.
25. Adam, W.; Boland, W.; Hartmann-Schreier, J.; Humpf, H. U.; Lazarus, M.; Saffert, A.; Saha-Moller, C. R.; Schreier, P. *Journal of the American Chemical Society* **1998**, *120*, 11044-11048.
26. Yin, M.; Baker, G. L. *Macromolecules* **1999**, *32*, 7711-7718.
27. Witzke, D. R.; Narayan, R.; Kolstad, J. J. *Macromolecules* **1997**, *30*, 7075-7085.
28. Dubois, P.; Jacobs, C.; Jerome, R.; Teyssie, P. *Macromolecules* **1991**, *24*, 2266-2270.

Chapter 13

An Integrated Process for the Production of Chemicals from Biologically Derived Succinic Acid

Nhuan Nghiem[1], Brian H. Davison[1], Mark I. Donnelly[2], Shih-Perng Tsai[2], and John G. Frye[3]

[1]Oak Ridge National Laboratory, One Bethel Valley Road, Oak Ridge, TN 37831
[2]Argonne National Laboratory, 9700 South Cass Avenue, Argonne, IL 60439
[3]Pacific Northwest National Laboratory, 902 Batelle Boulevard, Richland, WA 99352

An integrated process was developed for the production of industrial chemicals such as 1,4-butanediol, tetrahydrofuran, γ-butyrolactone, N-methyl pyrrolidinone, and 2-pyrrolidinone, from corn-derived glucose. The first step of the process is the production of succinic acid by fermentation of glucose using a new *Escherichia coli* strain (ATCC 202021), which over-produces succinic acid under anaerobic conditions. The fermentation process consists of two stages. In the first stage, the cells are grown under aerobic conditions until a critical cell mass is achieved. In the second stage, anaerobic conditions are applied to begin succinic acid production. This fermentation process has been successfully scaled up in a 500-liter fermenter. The succinic acid produced in the fermentation process is recovered by a two-stage desalting and water-splitting electrodialysis. The purified succinic acid then is catalytically converted to the final desired products. Using 1,4-butanediol as an example, an economic analysis shows that this integrated process can compete favorably with the currently used chemical processes.

Introduction

This project grew out of the Alternative Feedstocks Program of the U.S. Department of Energy (DOE), which directed a team of national laboratories to determine and develop alternative commodity chemical platforms based on renewable feedstocks. After several industrial workshops and reviews, and initial economic estimates by the National Renewable Energy Laboratory (NREL), succinic acid was chosen as a target platform chemical. In this exploratory effort, Oak Ridge National Laboratory (ORNL) focused on fermentation, Argonne National Laboratory (ANL) on metabolic engineering and membrane separations, Pacific Northwest National Laboratory (PNNL) on downstream catalytic conversion, and NREL on economic analysis.

Succinic acid is a chemical which has found applications in many areas; the most important ones being food additives, soldering fluxes, and pharmaceutical products *(1)*. The conversion of succinic acid to industrially important chemicals such as 1,4-butanediol (BDO), tetrahydrofuran (THF), γ-butyrolactone (GBL), N-methyl pyrrolidinone (NMP), and 2-pyrrolidinone (2P) recently has been made possible by the development of a number of catalytic processes *(2,3)*. This new development is expected to considerably expand the market for succinic acid.

Succinic acid currently is manufactured by chemical processes *(1)*. Many attempts have been made to develop a fermentation process for the production of succinic acid from renewable feedstocks such as corn-derived glucose. A number of patents have been issued on the microbial production of succinic acid *(4-7)*. However, none of these has been applied toward the commercial production of succinic acid.

In the search for a succinic acid producer, which is suitable for commercial application, the bacterium *Escherichia coli* has drawn significantly attention recently *(8,9)*. The probable reasons for interest in this organism are: 1. *E. coli* is a facultative microbe and relatively is easy to handle. It also has been known to be a rather robust organism; 2. *E. coli* has been used in many industrial fermentation processes, especially for the manufacture of biopharmaceuticals, and therefore, there exists a wealth of experience in large-scale application; and 3. The biochemistry and molecular biology of this organism are well known, therefore, improvement of the production strains is not a highly complicated task.

An *E. coli* mutant capable of producing succinic acid as the major product from glucose recently was discovered at ANL. The strain was initially called AFP111 and later was designated 202021 by the American Type Culture Collection (ATCC) in Rockville, MD. Realizing the potential of this strain, we have developed a fermentation process using an inexpensive industrial medium. A two-stage desalting and water-splitting electrodialysis process also has been developed for purification of the succinic acid product. This process concentrates, purifies, and acidifies the succinic acid, allowing the base to be recycled to the fermenter, where it is used for neutralization. New catalysts also have been developed for conversion of the purified succinic acid to BDO, THF, GBL, NMP and 2P. Figure 1 is a schematic diagram of the integrated process for the production of these chemicals from glucose. This chapter will discuss the four main areas of the process, which are strain development, fermentation, recovery/purification, and catalytic conversion.

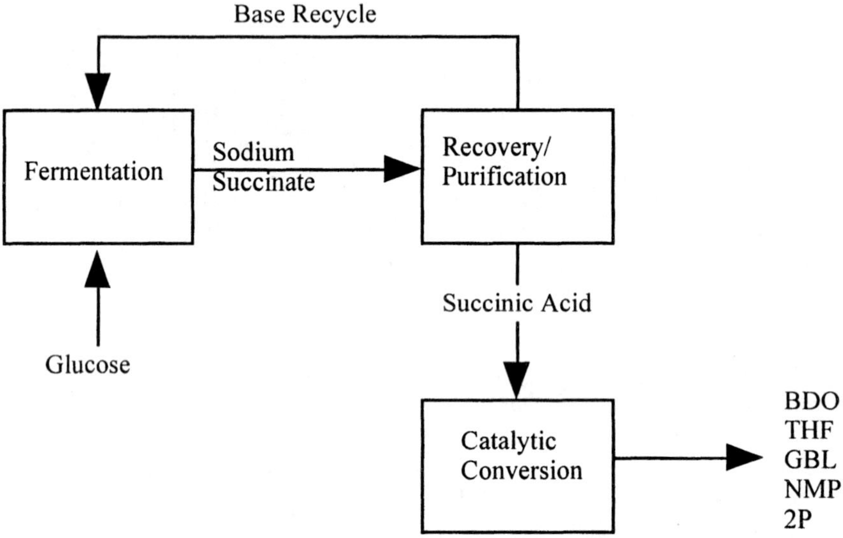

Figure 1. Schematic diagram of the integrated process for the production of chemicals from corn-derived glucose. Sodium succinate is shown for illustration purpose only. Different succinate salts will form depending on the bases used for pH control.

Strain Development

E. coli normally generates only a small amount of succinic acid from the fermentation of glucose, instead producing a mixture of products consisting primarily of acetic acid, formic acid and ethanol together with lower amounts of lactic and succinic acids *(10,11,12)*. The pathway is branched in that phosphoenolpyruvate is partitioned to final products that differ in their oxidation states. The proportion of products is balanced so that the NADH generated in glycolysis is fully reoxidized, allowing metabolism to continue *(12)*. Blocking branches of the normal fermentative pathways or introducing critical branch point enzymes *(13,14)* can alter the distribution of fermentation products.

Earlier, we improved the production of succinate in *E. coli* by overexpressing phosphoenolpyruvate carboxylase or malic enzyme in an appropriate host strain *(15,16)*. More recently, we described a mutant strain of *E. coli*, designated AFP111, that ferments glucose to succinic acid, acetic acid and ethanol in a molar stoichiometry of 2:1:1 *(8)*. AFP111 arose by a spontaneous chromosomal mutation of strain NZN111, a derivative of the wild type W1485 that is unable to ferment glucose due to inactivation of the genes for pyruvate:formate lyase and the fermentative lactate dehydrogenase *(17)*. The fermentation products formed from glucose by AFP111 and its lineage are illustrated in Figure 2. Quantitative analyses of the products formed by AFP111 show that approximately 1 mol of succinate is formed per mol of glucose consumed, with approximately 0.5 mol each of ethanol and acetate. These results are given in Table I, which describes the carbon and energy balances on a molecular basis. The fermentation is tightly coupled, as indicated by the recovery of carbon and the electron balance.

In the presence of exogenous hydrogen, the redox balance is altered; more succinate is produced, and less acetate must be formed to provide electrons for the reduction of PEP to succinate *(18)*. These results indicate that exogenous hydrogen can provide reducing equivalents to AFP111. Note that under these conditions AFP111 affects a net consumption of carbon dioxide, as indicated by the overall carbon balance (total recovery of 6.8 mol carbon atoms per mol glucose consumed). The positive redox balance reflects the input of electrons from hydrogen, which was not measured in these experiments. These experiments were performed in sealed serum tubes under 2 atm total pressure of a 1:1 mixture of H_2:CO_2. The hydrogen effect also occurs in the fermenter sparged with the same mixture of CO_2 and H_2 at 1.2. atm total pressure (see the Fermentation Process Development section below).

Recently, we have identified the mutation in AFP111 that allows it to ferment glucose as a lesion in the *ptsG* gene *(19)*. Introduction of a functional *ptsG* gene into AFP111 restored its parental, non-fermenting phenotype, and transduction of a mutant *ptsG* allele into NZN111 allowed it to ferment glucose, yielding principally succinic acid. The *ptsG* gene encodes the integral membrane permease of the phosphoenolpyruvate:glucose phosphotransferase system, and has both transport and regulatory functions. The enzymological and physiological characteristics of AFP111 were consistent with a mutation in this protein. Introduction of a mutant *ptsG* allele into other strains of *E. coli* altered their metabolism as well, resulting in the formation of succinic acid as the major product *(19)*.

A patent recently was issued on strain AFP111 *(18)*. This strain currently is designated ATCC 202021.

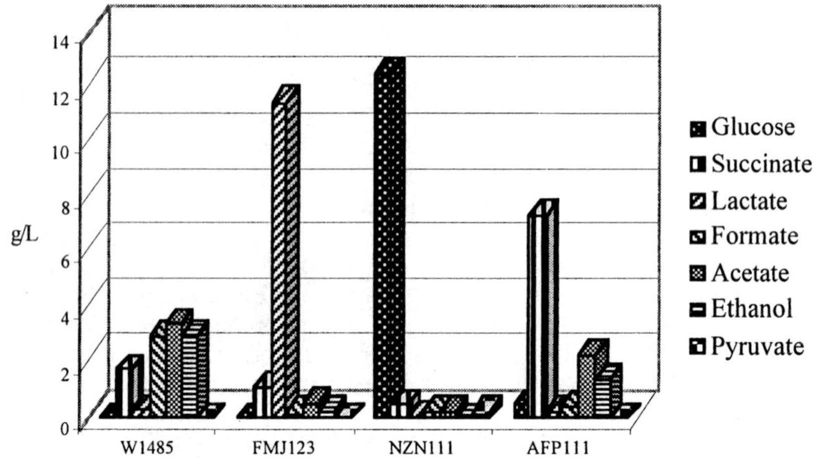

Figure 2. Products formed from the fermentation of glucose by the lineage of AFP111

Table I. Carbon and electron balances

	Products				
Atmosphere	*Succinate*	*Lactate*	*Acetate*	*Ethanol*	*Total*
CO_2					
mol/mol glucose	0.93	0.0	0.53	0.48	1.94
mol C atoms	3.72	0.0	1.06	0.96	5.74
O/R state	+0.93	0	0	-0.96	-0.03
H_2:CO_2					
mol/mol glucose	1.37	0.0	0.16	0.50	2.03
mol C atoms	5.48	0.0	0.32	1.00	6.80
O/R state	+1.37	0	0	-1.00	+0.37

O/R values are oxidation state times molarity. Expected totals for C atoms and O/R state are 6 and 0. Higher value in the presence of H2 indicate uptake of additional carbon and reducing equivalents, presumably from CO_2 and H_2.

Fermentation Process Development

Upon realizing the potential of strain ATCC 202021, an effort was launched to develop a fermentation process using this strain for commercial production of succinic acid. The developed fermentation process consists of two stages. In the first stage, the organism is grown under aerobic conditions until a critical cell mass is achieved. The second stage then begins with the discontinuation of aeration to provide anaerobic conditions in the fermenter. Succinic acid production starts immediately at high rates following the exhaustion of dissolved oxygen in the medium. This process recently was granted a patent *(20)*.

The fermentation process was first tested in a nutrient-rich medium, which contained 10 g/L Bacto-tryptone, 5 g/L Bacto-yeast extract, 1.4 g/L K_2HPO_4, 0.6 g/L KH_2PO_4, 2 g/L $(NH_4)_2SO_4$, 0.2 g/L $MgSO_4$, and 10 g/L NaCl. In these experiments, which were performed in a one-liter Virtis fermenter, the pH was controlled at 7.0 with a NH_4OH solution. The cells were grown aerobically for 24 hours before the air was turned off to begin succinic acid production. A 50% (w/v) glucose feed solution was continuously pumped into the fermenter. The speed of the feed pump was monitored and adjusted to maintain glucose concentration below 10 g/L. In a representative experiment, succinic acid concentration reached 30 g/L at 150 h at a yield of 0.63 g succinic acid/g glucose consumed.

After discussion of the results with the industrial partner, Applied CarboChemicals, Inc. (ACC), it was decided to develop an inexpensive fermentation medium for the fermentation process. Thus both tryptone and yeast extract were replaced with light steep water (LSW) in the next experiments performed at ORNL. A parallel effort was made at ANL to investigate the replacement of tryptone and yeast extract with corn steep solids (CSS). LSW was used at 25% (v/v) at ORNL and CSS at 2.5% (w/v) at ANL. In these experiments, the concentrations of the other ingredients were the same as those in the nutrient-rich medium. Very good agreement was obtained for the fermentation results generated at the two laboratories. Optimization of the fermentation process parameters then was performed in the LSW and CSS media in bench-scale fermenters. The parameters included the type of bases used for pH control, pH, and length of the aerobic stage. The following results were obtained:

- among the four bases used for pH control, which were sodium carbonate, ammonium carbonate, sodium hydroxide, and ammonium hydroxide, sodium carbonate gave the best succinic acid production
- the optimum pH was found to be between 6.25 and 6.5
- the length of the aeration stage should be from 4 to 8 hours

Figure 3 shows the results of a representative fermentation experiment using the CSS medium. In this particular experiment, the pH was maintained at 6.25 with NH_4OH. A succinic acid concentration of 54 g/L was obtained when the experiment was terminated at 144 h. The yield was 0.71 g succinic acid/g glucose consumed. A productivity of 0.63 g succinic acid/L-h was obtained during the linear production phase (from 6 h to 79 h). An experiment using the LSW medium under the same conditions yielded similar results.

The effect of exogenous hydrogen on succinic acid production observed in sealed serum tubes also was examined in a bench-scale fermenter. Figure 4 compares the results of this experiment and those of a control experiment. In hydrogen experiment, the fermenter was sparged with a 1:1 mixture of CO_2 and H_2 at 1.2 atm total pressure. In the control experiment, the hydrogen was replaced by nitrogen. Under these conditions, the yield of succinic acid increased from 0.63 to 0.90 g succinic acid/g glucose consumed in the presence of hydrogen, and the ratio of succinate to acetate increased from 11 to 25.

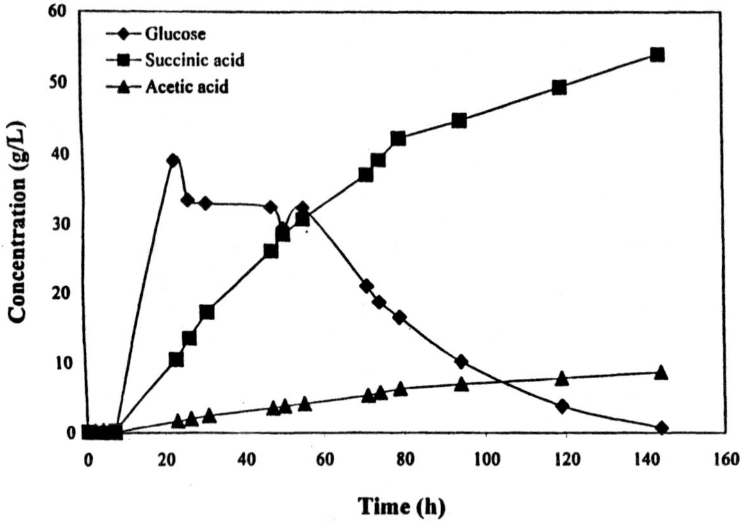

Figure 3. Concentration profiles of glucose, succinate, and acetate in a bench scale fementation experiment.

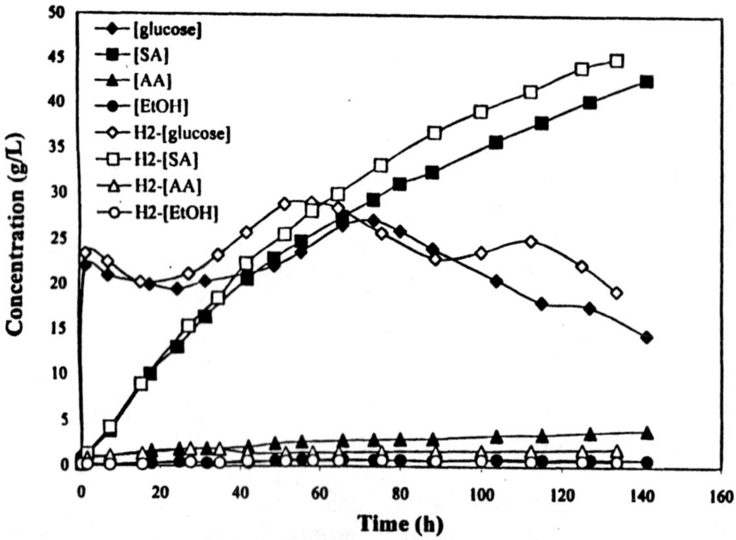

Figure 4. The effect of exogenous hydrogen on succinic acid production observed in a bench-scale fermenter

The developed fermentation process was scaled up in a 75-liter and then in a 500-liter fermenter. The succinic acid concentration profiles are shown in Figures 5 and 6. In these experiments, the LSW medium was used and the pH was maintained at 7.0 by Na_2CO_3. The results validated the fermentation process developed in bench-scale fermenters. They also showed a high degree of reproducibility. In experiment 97/15 in the 500-liter fermenter, the pH controller malfunctioned some time after 24 h and emptied the entire base reservoir into the fermenter. At 44 h, the pH was 7.6. An H_2SO_4 solution was pumped into the fermenter to bring the pH down. At 50 h, the pH was brought back down to 7.0. After a short period of no activity, succinic acid production started again. Although the final succinic acid concentration in this experiment was slightly lower than that in the other two experiments, the results clearly showed the robustness of the organism and its ability to withstand a common fermentation process upset condition such as the one described.

Product Recovery and Purification

The main objective of this development was to produce a purified succinic acid product for the catalytic conversion and a solution of ammonium succinate for a new application. The process used was a two-stage desalting and water-splitting electrodialysis process. This double-dialysis process avoids the generation of large quantities of salt wastes, which is a common problem in the recovery of fermentation-derived organic acid such as citric acid by the gypsum process *(21)*. The actual fermentation broths from the 75-liter and 500-liter scale-up experiments performed at ORNL were sent to ANL for use in the development of the product recovery and purification process.

Purification Processing Train for Ammonium Succinate Production

The purification processing train for ammonium succinate production consists of the following sequence of unit operations:

1. Microfiltration: removing cells and other suspended solids from the fermentation broth.
2. Nanofiltration: concentrating ammonium succinate and removing low-molecular-weight nonionic and monovalent impurities (e.g., sugars and acetate)
3. Desalting electrodialysis: further concentrating and purifying the succinate salt; removing small- and large-molecular-weight nonionic or weakly ionic species, such as sugars, alcohol (if any), amino acids, peptides, and proteins.

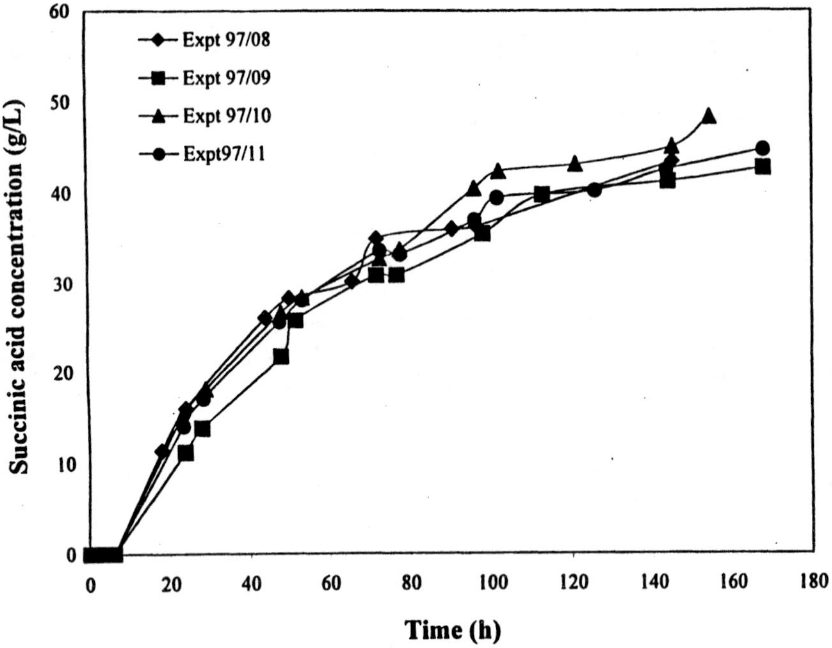

Figure 5. Succinic acid production in 75-liter fermenter

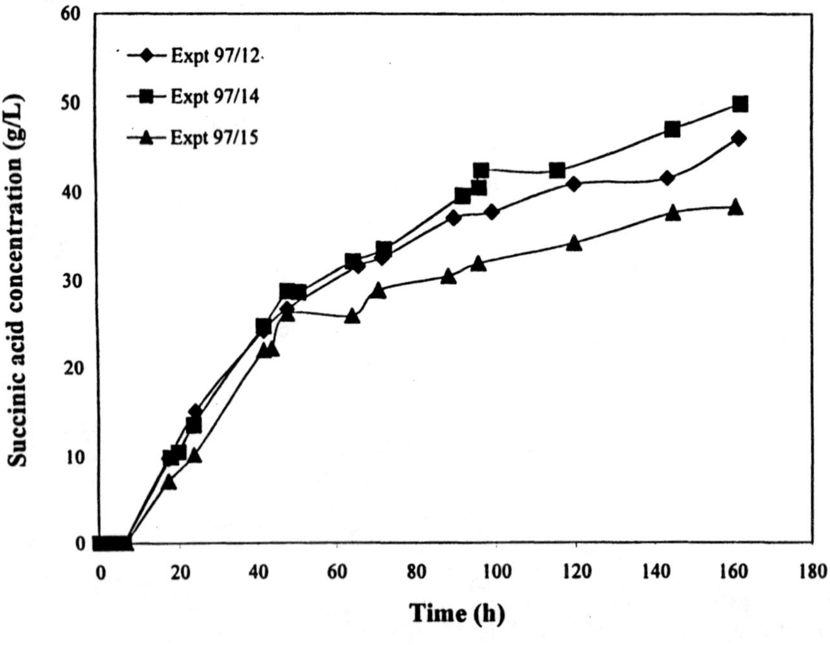

Figure 6. Succinic acid production in 500-liter fermenter

4. Activated carbon adsorption polishing: removing remaining sugars and proteins.

A laboratory system equipped with a Desalination System DK2540 spiral wound module was used for nanofiltration experiments. This module has about 1.8 m^2 of membrane area. This membrane has a molecular weight cutoff of about 150-300 and is reported to have high rejection of divalent anions in the presence of monovalent anions. Nanofiltration experiments were initially carried out using synthetic succinate broths. The membrane successfully permeated acetate and rejected succinate. Succinate rejection was found to increase with retentate pressure and was 97% at 400 psig, 94% at 300 psig, and 84% at 200 psig. Using a sequential nanofiltration and dialfiltration scheme, about 89% of the succinate in the original feed solution was recovered and 77% of the acetate was removed.

Production of Purified Succinic Acid

The fermentation broths were processed by nanofiltration to remove low-molecular-weight impurities and then by desalting electrodialysis to remove higher-molecular-weight impurities and to concentrate the succinate, followed by water-splitting electrodialysis to convert the ionized succinic acid to undissociated succinic acid. All processing steps proceeded smoothly, except for a membrane fouling incident in one of the desalting experiments. In that experiment, severe membrane fouling caused the run to be terminated. Upon examination of the interior of the membrane stack, the fouling appeared to be deposition of denatured (and insoluble) proteins (from the nutrient source) on the membrane surface. After cleaning using the established protocols, the membranes were reused successfully in subsequent runs, suggesting the fouling was reversible. Therefore, a two-prong approach to control membrane fouling was adopted, which consisted of reducing the protein contents in the broth prior to desalting electrodialysis and cleaning-in-place. In later runs, raising the pH of the broth before microfiltration was found to effectively cause a significant fraction of proteins to denature. The denatured proteins were then removed by microfiltration. This pretreatment reduced the membrane fouling propensity for nanofiltration and desalting electrodialysis. It also appeared to result in higher fluxes for the nanofiltration step.

The water-splitting electrodialysis products still contained small amounts of impurities, and aliquots of them were further purified by using different polishing schemes to evaluate the efficacy of each scheme. Table II shows the impurity profiles of purified succinic acid products obtained from different polishing schemes.

Table II. Impurity profiles of purified succinic acid products obtained from different polishing schemes.

Sample	TRS	Proteins	Sulfate	Lactic acid
SA-9801	<0.01	0.45	<0.01	0.77
SA-9802	<0.01	0.62	2.34	0.50
SA-9803	<0.01	0.01	0.56	<0.01

TRS: total reducing sugars. All impurity concentrations are expressed as weight % of succinic acid (wt % SA). Anion-exchange polishing was used only for SA-9801. Crystallization was used only for SA-9803.

Catalytic Conversion of Succinic Acid

Preliminary investigations at PNNL were on the production of BDO, THF, GBL, NMP and 2P from succinate. Catalysts were identified for the selective reduction of succinic acid to GBL in the presence of fermentative by-product acetic acid. While high GBL yields were successfully demonstrated, catalyst productivities were still below commercially attractive rates. Subsequently, a very promising catalyst and set of processing conditions had been found that was capable of converting GBL to 1,4 BDO at high rates and at nearly 100% selectivity. This was intended to be part of a two-step process (succinic acid to GBL to 1,4 BDO). Also, several catalyst systems had been identified that were capable of converting succinic acid to a mixture of GBL, 1,4 BDO, and THF in a single step.

Catalysts and processing conditions for the production of 2P from succinimide or diammonium succinate also were investigated. A very promising catalytic and non-aqueous process was discovered for converting succinimide to a mixed product containing 2P and GBL. However, early attempts at preparing 2P via aqueous processing considerably lagged behind results reported in the patent literature. A few preliminary investigations had also been made into the production of NMP in single-step reactions. While several of these reactions yielded the desired NMP product, rates of production were exceptionally low.

We also explored business interest in the production of 2P and related derivatives. A biologically derived ammonium succinate or succinimide feedstock could compete with 2P and related derivatives produced by petrochemical routes. Key technical highlights in the area of catalyst and process development at PNNL follow.

Semi-batch reactor tests were used to map reaction pathways and to obtain rate data to guide rational catalyst selection for the production of 2P from aqueous ammonium succinate solutions. Rh/C catalysts appeared to offer the best compromise of reaction rate with the least amount of over-reduction of the 2P product (therefore displayed the highest product yields). 2P yields as high as 82%, and at high conversion rates (in excess of the targeted 1 g per hour per gram of catalyst) were achieved within a relatively short time, and were comparable to prior patent literature claims.

Flow reactor studies of the production of 2P from aqueous ammonium succinate solutions showed that the Rh/C catalysts that worked well in semi-batch tests rapidly deactivated in longer runs. Significant Rh metal sintering was observed in spent catalyst samples, which is thought to be largely responsible for the large losses of catalytic activity that were observed in the early flow reactor studies. New promoted Rh/C catalysts were developed and found to be substantially more active that the previous Rh/C catalysts (60% higher initial catalytic activity than that displayed by the Rh/C catalysts and containing an equivalent amount of Rh metal). These catalysts were termed "textured" catalysts.

Significant improvements in the conversion rates of pristine succinic acid solutions to GBL were achieved with "textured" versions of Pd/C catalysts. In semi-batch tests at reaction temperatures of 225°C, productivities as high as ~ 1.7 g per hour per gram of catalyst were observed while displaying nearly 100% selectivity to the desired GBL product.

Process Economic Analysis

These results were obtained from a technical and economic analysis performed by NREL and by New Horizon, Inc., for this project. This has been reviewed by several industrial representatives, updated by ORNL and Energetics, Inc., and reviewed by A.D. Little. The calculations were based on a plant with a capacity of 105 million lbs/yr BDO. The capital cost of the conventional technology was higher than the new technology.

Below is a nonproprietary summary showing the comparison. The primary competitors with the biologically derived succinic acid (BDSA) process are the three currently available petrochemical-based processes. As indicated in Table III, the Reppe process, which has been described most extensively in published reports, has the lowest "published" cost among the competitors ($0.66/lb). The estimates for other processes are actually expected to be lower despite the published external cost reports. Economic estimates concerning competing technologies are from SRI International reports *(22)*; those relating to BDSA are from a detailed process flowsheet, based on a plant producing 105 million lb of BDO per year. Pollution and energy estimates are from the SRI reports and from a life-cycle analysis report by PNNL *(23)*.

Table III. Comparison of BDSA process with petrochemical-based processes

Process	*BDSA*	*Reppe*	*Kuraray/ ARCO*	*Davy-McKee*
Raw material	Corn sugar	Petroleum, natural gas	Syngas	Butane
Estimated production cost ($/lb BDO)	0.43	0.66	1.14	0.89
Raw material cost ($/lb BDO)	0.20	0.42	0.99	0.67
Energy use (Btu/lb BDO)	36,000	42,000	39,000*	-------

*Estimated from utilities cost.

As shown in Table III, the BDSA process offers significant advantages in two primary areas. First, the process has a significantly lower estimated production cost for BDO, succinic acid, and several other commodity chemicals, as well as lower energy consumption during production. Current corn production could easily supply the 100 to 200 hundred million pounds of corn sugar required for each major chemical plant, and use of this feedstock would decrease petroleum consumption and imports. Further analysis by ACC and New Horizon of the production cost for succinic acid in our fermentation process has included several process modifications to further decrease costs. The estimated costs in 1998 were $0.20/lb sodium succinate. However, the petrochemical routes to BDO have also had further improvements, increased capacity and presumably lowered the target price.

Acknowledgement

This project was supported by the U.S. Department of Energy's Office of Industrial Technology (DOE-OIT). The project could never have succeeded without the relentless efforts of the following people: Gloria Kulesa, Doug Faulkner, and Linda Schilling of DOE-OIT, Eric O'Connor and Mike Griffin of ACC, Ranjini Chatterjee, Cynthia Sanville-Millard, Lucy Stols, Kathleen Champion, May Wu, Mike Henry, and Tony Fracaro of ANL, Ron Landucci, Gene Petersen, Joan Pelligrino, and Mark Ruth of NREL, Grady Vanderhoofen, Bruce Suttle, Gerald Richardson, John Parrott, Miguel Rodriguez, Jr., Cathy McKeown, and Maria Blanco-Riverra of ORNL, Todd Werpy, Yong Wang, Alan H. Zacher, Todd R. Hart, and Sarah Burton of PNNL. Eric Johnson at the University of Illinois Urbana-Champaign Biotechnology Center Fermentation

Facility performed fermentations in the presence of hydrogen. A. E. Staley in Loudon, Tennessee generously provided the light steep water.

References

1. Fumagalli, C. In *Kirk-Othmer Encyclopedia of Chemical Technology;* Kroschwitz, J. I. and Howe-Grant, M., Eds.; Wiley: New York, NY, 1997: Vol. 22, pp 1072-1102.
2. Rao, V. N. M. U.S. Patent 4,782,167, 1988.
3. Mabry, M. A. U.S. Patent 4,550,185, 1985.
4. Ling, L. B. U.S. Patent 4,877,731, 1989.
5. Datta, R. U.S. Patent 5,143,833, 1992.
6. Glassner, D. A. U.S. Patent 5,143,834, 1992.
7. Guettler, M. V. U.S. Patent 5,573,931, 1996.
8. Donnelly, M. I.; Millard, C. S.; Clark, D. P.; Chen, M. J.; Rathke, J. W. *Appl. Biochem. Biotechnol.* **1998**, *70/72,* 187-198.
9. Wang, X.; Gong, C. S.; Tsao, G. T. *Appl. Biochem. Biotechnol.* **1998**, *70/72,* 919-928.
10. Blackwood, A. C.; Neish, A. C.; Ledingham, G. A. *J. Bacteriol.* **1956**, *72,* 497-499.
11. Bock, A. and Sawers, G. In *Escherichia coli and Salmonella typhimurium;* Neidhardt, F. C., Ed; Washington: DC, 1996, pp 262-282.
12. Clark, D. P. *FEMS Microbiol. Rev.* **1989**, *63,* 223-234.
13. Gupta, S.; Clark, D. P. *J. Bacteriol.* **1989**, *171,* 3650-3655.
14. Ingram, L. O.; Conway T.; Clark, D. P.; Sewell G. W.; Preston, J. F. *Appl. Environ. Microbiol.* **1987**, *53,* 2420-2425.
15. Millard, C. S.; Chao, Y.-P.; Liao, J. C.; Donnelly, M. I. *Appl. Environ. Microbiol.* **1996**, *62,* 1808-1810.
16. Stols, L.; Donnelly, M. I. *Appl. Environ. Microbiol* **1997**, *63*, 2695-2701.
17. Bunch, P. K.; Mat-Jan, F.; Lee, N.; Clark, D. P. *Microbiol.* **1997** *143,* 187-195.
18. Donnelly, M. I. U.S. Patent 5,770,435, 1998.
19. Chatterjee, R.; Millard, C. S.; Champion, K.; Clark, D. P.; Donnelly, M. I. *Appl. Environ. Microbiol.* (submitted).
20. Nghiem, N. P. U.S. Patent 5,869,301, 1999.
21. Atkinson, B. and Mavituna, F. *Biochemical Engineering and Biotechnology Handbook,* Stockton Press: New York, NY, 1991; 1173-1174.
22. *SRI International PEP Yearbook,* 1993; Vol. 1, 80-83.
23. *Streamlined Life-Cycle Assessment of 1,4-Butanediol Produced for Petroleum Feedstocks Versus Bio-Derived Feedstocks,* PNNL-11213, Pacific Northwest National Laboratory: Richland, WA, 1996.

Chapter 14

Use of Model Compounds to Study the Reactivity and Cross-Linking of Natural Phenolics

Stephen S. Kelley, Xiang-Ming Wang, Michele D. Myers, and Mark F. Davis

Chemical Bioenergy Systems, National Renewable Energy Laboratory, 1617 Cole Boulevard, Golden, CO 80401–3393

Fast pyrolysis of biomass produces a phenol-rich oil suitable for incorporation into phenol-formaldehyde (PF) resins. To evaluate the reactivity and network formation characteristics of the compounds typically found in these oils, a series of phenolic model compounds was reacted with formaldehyde under conditions typically used to prepare PF resins. This study indicates that the substituted phenolics commonly found in pyrolysis oils are more reactive than phenol. It also showed that the network formation process for these complex phenolic mixtures follows the predictions of the simple statistical approaches developed by Flory *(23)* and Stockmayer *(24)*. These results show that the substituted phenolics commonly found in pyrolysis oils will be chemically bonded into the polymer network and that under the proper reaction conditions a highly stable network will be formed.

Introduction

Natural Phenolics in Phenol Formaldehyde Resins

Phenol-formaldehyde (PF) resins are widely used as thermosetting adhesives for exterior-grade wood composites. Since phenol is traditionally derived from petroleum-based products, the production of PF resins from such phenol is relatively expensive and subject to changes in the price of petroleum. One approach for reducing the cost of PF resins is to replace the petroleum-based phenol with phenols derived from renewable materials. Numerous studies

(1–4) have focused on using renewables such as lignin or tannins to replace phenol. However, both materials have limitations—including the relatively low reactivity of lignins and the relatively high cost of tannins—that have prevented their widespread use in North America. The relatively low reactivity of lignin is related to a low number of free phenolic hydroxyl groups and a lack of "free" ortho- and para- reactive sites on the aromatic ring. The high cost of tannins is related to their low yield from bark.

Other natural phenolics are derived from biomass pyrolysis. These oils are chemically complex but have more reactive sites on an individual phenolic ring and can be recovered in relatively high yields *(5–9)*. Thus, the use of phenolic-rich oils isolated from biomass pyrolysis offers a way to overcome the limitations of lignins and tannins as replacements for phenol.

Fast pyrolysis of biomass can produce phenolic-rich oils in high yields, 40%–50% based on the weight of the dry feed *(5)*. These oils can be used to replace some of the petroleum-derived phenol used to prepare PF resins. However, to effectively replace the monomeric phenol, the phenolic oils must be chemically incorporated into the PF resin network under typical resin synthesis conditions; they cannot simply act as fillers or extenders.

Since these phenolic-rich oils are complex mixtures of more than 100 distinct compounds, they are difficult to study. We selected a series of well-defined model compounds that can emulate the behavior of the phenolic-rich oils. The two reaction characteristics of interest are the inherent reactivity of these substituted phenolics with formaldehyde and their assembly into crosslinked polymers (gelation).

Phenolic Model Compounds

Studies on the reactivity of phenolic model compounds are useful in understanding the reactivity of biomass-derived phenolic compounds with formaldehyde under various reaction conditions. Studies over a 40-year period have shown that the reaction of phenol with formaldehyde followed second-order kinetics *(10,11)*. A series of kinetic studies of the reactions of 3-methylphenol, 2-methylphenol, and 2,5-dimethylphenol with formaldehyde showed that the reaction between formaldehyde and phenol model compounds also followed second-order kinetics and that the reaction rates increased with increasing temperature and pH *(12–14)*.

^{13}C-NMR spectroscopy was used extensively to investigate the chemical structures of phenolic resins *(11,15–18)* and related model compounds *(17–19)*. These studies provided useful information on the complex chemical structure of PF resins. ^{13}C-NMR spectroscopy was also used to study the chemical structure and to determine the copolymer composition of novolac resins prepared from 2-methylphenol, 3-methylphenol or 4-methylphenol *(20,21)* or from mixtures of 3-methylphenol and 4-methylphenol *(22)*.

Gelation of Phenolic Mixtures

One of the earliest approaches to studying the crosslinking reactions of polymer networks is based on a simple statistical analysis of the polymer system *(23,24)*. This work included three important assumptions: 1) all functional groups of the same type are equally reactive, 2) all functional groups react independently of one another, and 3) no intramolecular reactions occur. More elaborate approaches consider the reactivity differences for various sites on the monomers and changes in reactivity of a site as the reaction proceeds *(25–29)*.

The simple statistical approach considers three molecular features of the crosslinking system. First is the ratio of reactive sites. Since phenol has three reactive sites and formaldehyde has only two, this ratio differs from the mole ratio. The second molecular feature is the number of reactive sites on a single molecule. If this number is greater than three, the molecule can act as a crosslinking site for more than three growing chain ends. This feature can be very important for natural phenolics that may have 10 or more reactive sites on a single molecule. The third molecular feature is the number of molecules that cannot serve as crosslinking sites. If the molecule has only two reactive sites, it will allow chain growth to continue and will increase the molecular weight between the branch points. If the molecule has only one reactive site, it will terminate the growth of that chain, resulting in a defect in the network.

Although the assumptions of the simple statistical approach of Flory *(23)* and Stockmayer *(24)* do not hold for PF resin, it was used to model the gelation behavior of these model systems because the real pyrolysis oil systems are so complex that the more elaborate modeling approaches could not be applied to these systems.

Materials and Methods

Experimental Approach

To better understand how the complex phenolics produced by biomass pyrolysis can be chemically incorporated into the PF resin network, we used a series of phenolic model compounds to study the reaction rate and the gelation processes under conditions typically used to synthesize PF resins. The model compounds (Figure 1) include 2-methoxy-4-methylphenol, 4-methylcatechol, 2-methylphenol (o-cresol), 4-methylphenol (p-cresol), phenol, 4,4'-isopropylidenediphenol (Bisphenol A), and trisphenol. Figure 1 also shows the number of reactive sites (f) on each model compound.

In the kinetics studies, the model compounds were reacted with formaldehyde under a variety of reaction conditions that are typically used to synthesize commercial PF resin. For the kinetics studies the variables included reaction temperature and time, molar ratio of formaldehyde to phenol (F/P), and

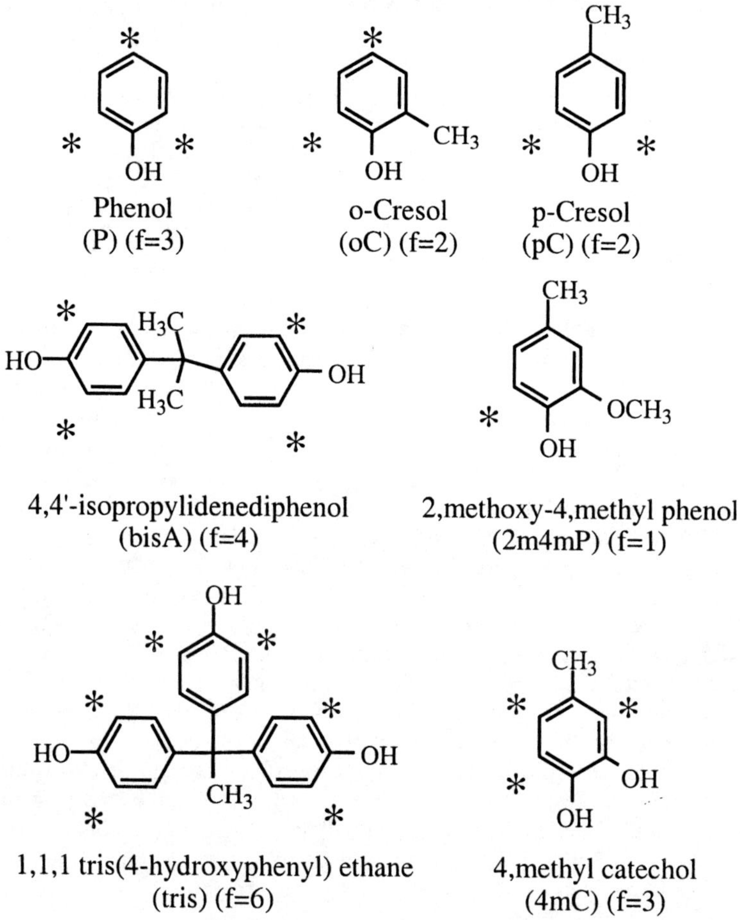

Figure 1. Model compounds used to represent biomass pyrolysis oils. The asterisks represent sites that can potentially reaction with formaldehyde under stand PF resin synthesis conditions.

molar ratio of sodium hydroxide to phenol (NaOH/P). The progress of the reaction was followed with ^{13}C-NMR by monitoring the decrease in the concentration of the starting material. The results of kinetics studies were then compared to that of phenol, which was also cooked with formaldehyde under the same conditions.

For the gelation studies, standard PF resin conditions were used. The effects of the ratio of reactive sites (phenol/formaldehyde), the average functionality of the phenolic system, and the amount of monofunctional phenolics were studied. Since the solids content and pH of all systems were the same, solution viscosity was used to monitor the system's average molecular weight.

Kinetic Studies

We obtained all chemicals from commercial sources and used them without further purification. Sodium hydroxide was used as a 50% aqueous solution.

In the kinetics experiments, the variables included reaction temperature (40°C, 60°C), NaOH/P mole ratio (0.05, 0.25), and F/P mole ratio (1.20, 1.80). To minimize concerns about secondary condensation reactions or solubility, we used a solids content of 20%. In the gelation experiments, the F/P ratio was varied from 0.60 to 1.40, and the NaOH/P mole ratio was held constant at 0.25.

All the kinetics experiments were conducted using a 500-mL three-neck round-bottomed flask equipped with a reflux condenser, a thermometer, and a stirrer. The temperature was controlled to ±1°C. First, the flask was charged with paraformaldehyde (95% solid content), phenol (89.1% aqueous solution) or model compound, 1,4-dioxane, and methanol. The sodium hydroxide (50% aqueous solution) and the required amount of water were then added slowly over 5 min while the solution was stirred. When the entire mixture became soluble (transparent), the first sample (5–10 mL) was taken and the heater was turned on. The initial sample taken at room temperature was used as the zero time sample. After reaching 40°C or 60°C, a sample was taken every 30 min for the first 2 h and every hour afterward until 8 h (the longest cooking time). To prevent any further chemical reaction, all samples were frozen immediately after being removed from the reactor.

Gelation Studies

These experiments were similar to the kinetics experiments. The reactions were run at 95°C, and the reaction time was started when the reaction reached 95°C. Samples were taken frequently and the viscosity was tracked by comparing the resin viscosity to that of Gartner-Holdt tubes. All samples were rapidly cooled and stored in a freezer until the viscosity could be measured. A Brookfield Dial Viscometer (model DV-1) was used to measure the solution viscosity of the samples at 25°C.

Results

Reactivity of Model Compounds

Conversion of ^{13}C-NMR intensity into concentrations

The reactivity of a compound, determined with ^{13}C-NMR, was defined as the loss of starting material as measured by the relative intensity of NMR peaks assigned to phenol or a model compound. The NMR spectra were not run under quantitative conditions because of the long ^{13}C relaxation times associated with the substituted phenolic carbons. The ^{13}C relaxation times were measured for 2-methoxy-4-methylphenol and the longest T_1 was measured to be 35 s for the C2 aromatic carbon. Using this T_1 value, we determined that, for the number of reaction times and conditions studied for phenol or each phenolic model compound, quantitative data would have taken too long to collect using the requirement of a delay time that is 5 times the longest T_1. To obtain quantitative data on the concentration of the phenolic model compound, 1,4-dioxane was added to the reaction mixture as an internal standard. NMR followed the change in intensity at the C5 position of each compound under each reaction condition. Although the NMR spectra were not run under quantitative conditions, the relative intensities of the C5/dioxane ratio should not change unless there is a change in the concentration of either compound.

Reaction kinetics of phenol and model compounds

For each compound, a plot of the logarithm of the product compound concentration and formaldehyde concentration (mole/L) versus the reaction time (s) shows a linear correlation under each reaction condition. This indicates that the reaction of phenol or model compounds with formaldehyde followed the second-order reaction. The overall rate expression is shown as follows:

$$kt = \frac{1}{a - b} \ln \left[\frac{b(a - x)}{a(b - x)}\right] \qquad [1]$$

where
k = second-order rate constant (L/mole-s)
a = initial concentration of phenol or phenolic model compound (mole/L)
b = initial concentration of formaldehyde (mole/L)
x = amount of phenol or phenolic model compound reacted (mole/L) at different time intervals (s)

The results showed that the reactivity of all model compounds was highest at 60°C with NaOH/P=0.25. Under other reaction conditions, the general order of reactivity (except for 4-methylcatechol at high temperature/low base) was high temperature/high base > high temperature/low base > low temperature/high base > low temperature/low base. These results indicate that increasing both reaction temperature and amount of base significantly enhances the reactivity of phenol and phenolic compounds with formaldehyde; however, the former seems more important. These results are consistent with the kinetic studies by Malhotra *et al.*, who found that the reaction of 2-methylphenol *(12)*, 3-methylphenol *(13)* or 2,5-dimethylphenol *(14)* with formaldehyde followed a second-order reaction under basic conditions. Also, the reaction rate increased when both temperature and pH were increased.

Under all reaction conditions, the reactivity of all three model compounds was higher than that of phenol (see Figure 2). Under the higher reaction temperature with higher levels of NaOH/P molar ratio and F/P molar ratio, for example, the reaction of 2-methoxy-4-methylphenol, 2-methylphenol, and 4-methylcatechol with formaldehyde was 3.5, 2.7, and 6.0 times faster than that of phenol. These results indicate that the pyrolysis oils will be highly reactive under traditional PF resin synthesis conditions and is likely to be chemically incorporated into the PF network.

Gelation Studies

Theoretical Approach

Resin chemists are familiar with the reaction sequences common to commercial condensation resins such as PF or urea formaldehyde resins. These sequences involve prolonged reaction times with little or no change in the viscosity of the resins, followed by a period when the viscosity increases rapidly. If unchecked by rapid cooling, changes in pH or ionic content, or the addition of solvent, this reaction will lead to the formation of a solid resin in the reaction kettle in a matter of minutes.

These condensation reactions can be modeled based on simple statistical approaches. One of the first approaches for predicting gelation or the gel point of complex mixtures was developed by Flory *(23)* and Stockmayer *(24)* (see equation 2).

$$X_n = \frac{f(1 - \rho + 1/r) + 2\rho}{f(1 - \rho + 1/r - 2p) + 2\rho} \qquad [2]$$

where

X_n = number average degree of polymerization

f = average functionality of the system or the sum of all phenolic sites divide by the number of reactive molecules

ρ = ratio of nonbranching phenolic reactive sites to the total number of phenolic reactive sites

r = ratio of phenol reactive sites to formaldehyde reactive sites

p = extent of reaction which is related to reaction time

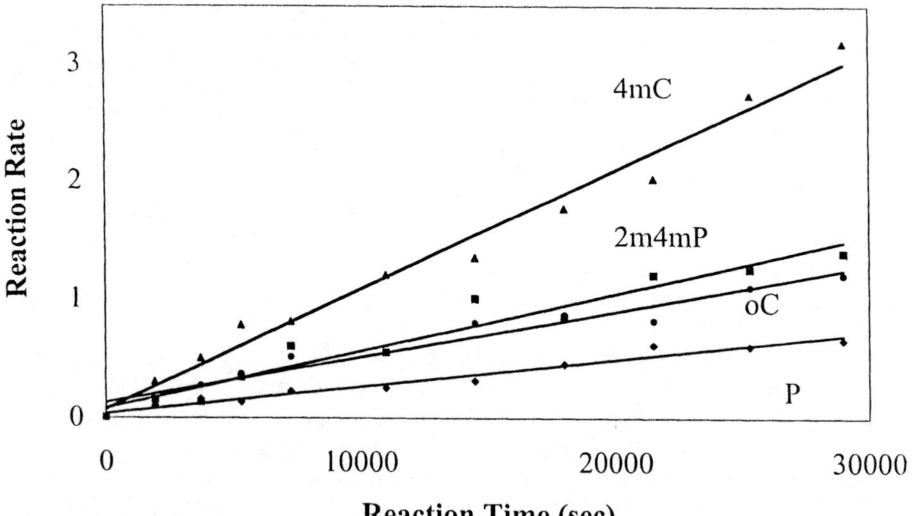

Figure 2. A comparison of the reactivity of phenol and substituted phenolic model compounds.

Equation 2 can be used to predict increases in the degree of polymerization (or molecular weight) as the extent of reaction (or reaction time increases). In an industrial environment this equation is expressed in terms of reaction time and solution viscosity. It predicts the well-known induction period with little increase in molecular weight or viscosity until a critical point is reached where the degree of polymerization or molecular weight raises rapidly. It also shows how changing the ratio of reactive groups, the number of reactive sites on a single phenolic molecule, and the concentration of nonbranching phenolic molecules can affect the polymerization reactions that lead to gelation.

The impact of the ratio of reactants is shown in Figure 3a. Increasing the ratio of the reactants (in this case the ratio of reactive phenolic sites to reactive formaldehyde sites) decreases the extent of reaction or time required to reach the gel point. The (r) value used in equation 2 is based on the ratio of reactive sites, not on the mole ratio of the components. Therefore, changing the type of phenolic used to prepare a resin, rather than simply replacing phenol with cresol or bisphenol A, requires that the ratios be recalculated. In systems that are deficient in phenolic reactive sites ($r<1.0$), the extent of reaction or reaction time must be quite high for gelation to occur. For systems with a substantial excess of phenolic sites ($r=1.4$) the reaction time required to reach the gel point can decrease more than 15% relative to a system with balanced stoichiometry. Changing the ratio of reactive sites has the largest impact on the time required to reach the gel point.

The number of reactive sites on a single phenolic molecule, i.e., the "functionality" of that molecule, can also have a dramatic impact on the gel point (Figure 3b). Increasing the functionality from phenol ($f=3$) to bisphenol A ($f=4$) to trisphenol ($f=6$) decreases the reaction time by about 15%. If the PF resins are made with phenolics that have higher functionalities, e.g., lignins or high molecular weight pyrolysis oils, these changes in functionality will reduce the cooking time needed to reach the gel point. Reductions in functionality are not common for resole type PF resins, since most are made only with phenol, although some novolac resins use substituted phenols that reduce the overall functionality of the system and increase cooking times. Equation 2 suggests that adding 33 mol% of ortho- or para-cresol to standard PF resin formulation increases the extent of reaction or the reaction time at the gel point by about 5%.

Finally, equation 2 considers the impact of adding non-branching phenolics, which include phenolics with two or even one reactive site, to the reaction mixture. Phenolics with two reactive sites will extend the molecular weight between the branch points; monofunctional molecules will terminate the growing chain end. Monofunctional molecules are particularly important because they will create a chain end or defect in the resin network. Although monofunctional molecules have a disastrous impact on the formation of high molecular weight linear polymers, they have minimal effect on crosslinking resins. The predicted impact of adding up to 40 mol% of non-branching molecules to a standard PF resin recipe is shown in Figure 3c. Although adding nonbranching molecules increases the extent of reaction or time required to reach

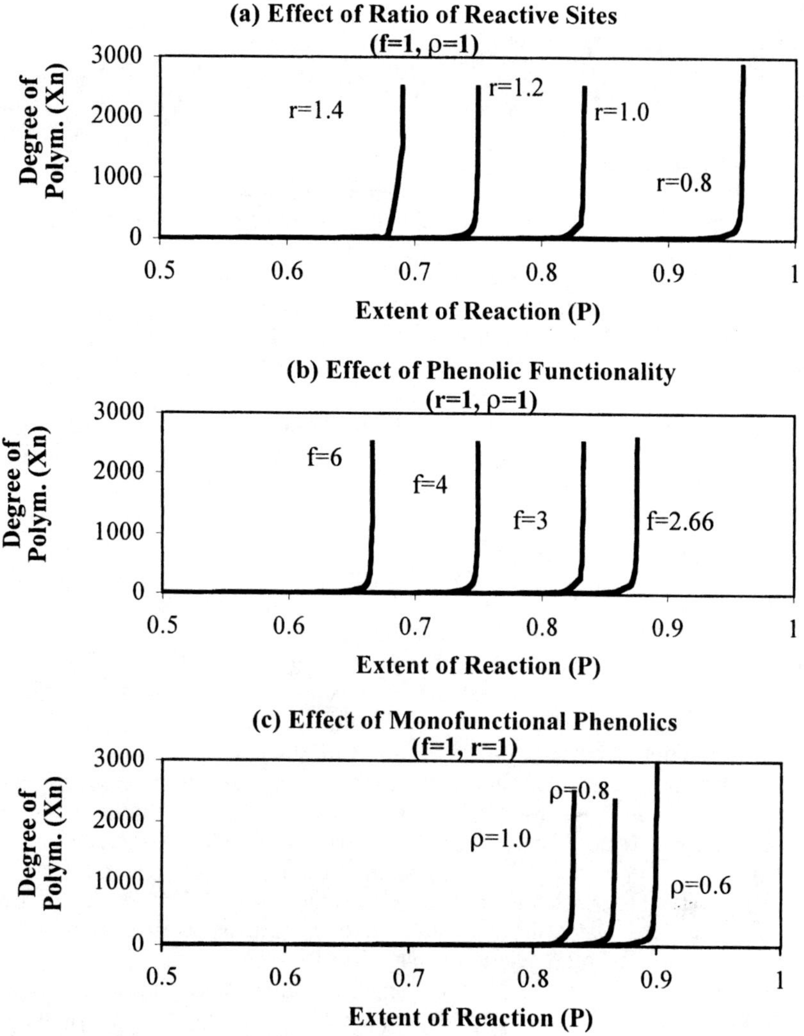

Figure 3. Changes in the degree of polymerization as a function of the extent of reaction calculated from equation 2; (a) changes in the ratio of phenol to formaldehyde reactive sites (r) for constant (f) and (ρ) (f=3, ρ=1), (b) changes in the functionality or number of reactive sites on a individual phenolic molecule (f) for constant (r) and (ρ) (r=1, ρ=1), and (c) changes in the amount of monofunctional phenolics (ρ) for constant (f) and (r) (f=1, r=1).

the gel point, the increase is only 2%-3% for every 10 mol% of monofunctional molecules added to the reaction.

Results from Model Compounds

To confirm the predictions from equation 2, a series of model compounds was used to create PF type resin formulations with varying ratios of the two reactive components, phenol functionalities and monofunctional phenol content. The solids content, ratio of phenolic hydroxyl groups to sodium hydroxide catalysts, and reaction temperature were kept constant. This allows for the rapid measurement of the solution viscosity, which can then be equated to the number average degree of polymerization, which is relatively difficult to measure. The extent of the reaction, which is also very difficult to measure accurately in these systems, was equated to the reaction time.

Four reactive systems were studied at varying ratios of the two reactive components. These reactive systems were phenol, bisphenol A, trisphenol, and a mixture of bisphenol A and o-cresol (results from the latter three systems are not shown). An example of the effects of increasing (r) on the reaction time needed to reach the gel point is shown in Figure 4a for phenol. Clearly, the time required to reach the gel point decreases as the ratio of the reactants increases, as predicted by equation 2. The gel time for the other phenolic systems also decreases as the ratio of reactants increases. These model systems show that changing the ratio of reactive sites has the same impact regardless of the functionality of the phenolic system. This is very important as one tries to apply the results from model compound studies to real phenol formaldehyde adhesives that may include complex mixtures of phenolics.

Similar data at a constant (r) value show that the gel time decreases as the functionality (f) of the phenolic system increases. Again this result is consistent with predictions from equation 2. As shown in Figure 4b, increasing the functionality of the phenolic component, moving from phenol (f=3) to bisphenol A (f=4) to trisphenol (f=6) decreases the reaction time needed to reach the gel point. This result shows how a small amount of material with a high functionality can affect the gelation behavior of PF resins.

Finally, the effects of nonbranching phenolics were investigated with several systems. Figure 4c shows the impact of adding different amounts of a monofunctional phenolic (2-methoxy-4-methylphenol) to the formulation. As predicted by equation 2, adding nonbranching phenolics increases the time required to reach the gel point.

In an effort to better understand the impact of difunctional and monofunctional phenolic molecules, four resins were prepared at two ρ values. The resins were formulated with mixtures of two branching molecules, bisphenol A and trisphenol, and a difunctional phenolic (o-cresol), a monofunctional phenolic (2-methoxyl-4-methylphenol) (see Figures 5a and 5b). Figure 4 shows the impact of adding both difunctional and monofunctional phenolic molecules to produce mixtures with varying ρ. By varying the ratio of

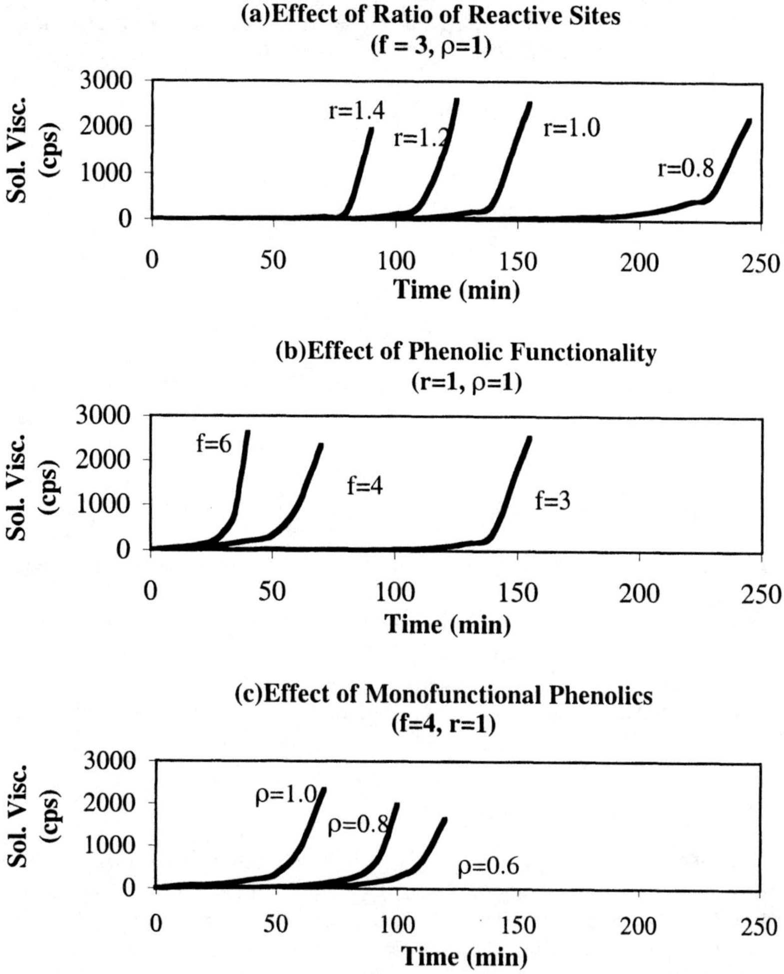

Figure 4. Changes in solution viscosity as a function of reaction time for simple model compound systems; (a) the effect of change the ratio of phenol to formaldehyde reactive sites (r) for constant (f) and (ρ) (f=3, ρ=1), (b) the effect of changing the functionality of the phenolic model compound (f) for constant (r) and (ρ) (r=1, ρ=1), and (c) the effects of adding a monofunctional phenolic to the reactive system (ρ) for constant (f) and (r) (f=1, r=1).

the four components these formulations were produced to have the same average functionality (f) and ratio of phenolic sites to formaldehyde sites (r) but differing amounts of monofunctional and difunctional molecules (ρ). According to equation 2 all four curves at each ρ level should follow the viscosity time profile. The viscosity curves are not sensitive to differences between difunctional (pC) and monofunctional (2m4mP) phenolic molecules. But clearly the bisphenol A (bis) and trisphenol (tris) systems differ.

Discussion

Reactivity of Model Phenolics

The data in Figure 2 clearly show that the phenolic model compounds studied here are more reactive than monomeric phenol under standard phenol formaldehyde reaction conditions. This is not surprising, since a wealth of literature shows the phenols substituted with electron donating groups are more reactive than unsubstituted phenol *(12-14)*. However, it is particularly important for these complex phenolic-rich pyrolysis oils. These results suggest that the pyrolysis oil phenolic will react rapidly under standard cooking conditions and become covalently bound into the polymer network. This means that the pyrolysis oils contribute to the strength of the network and do not simply act as fillers or extenders.

This also means that odorous components of the pyrolysis oils, many of which are phenolic *(30)*, are chemically bonded into the network with stable methylene type linkages. Thus, the concentration of unreacted phenolics will be greatly reduced during resin synthesis and the phenolics that are bonded into the network are not likely to be released during the product lifetime.

Network Formation of Model Phenolics

In the case of biomass pyrolysis oils, all three chemical features defined in equation 2 are important. To some extent the resin chemist can control the ratio of reactive sites (r). But in the case of phenolic-rich pyrolysis oils, determining the actual number of reactive phenolic sites is very difficult. The number of reactive phenolic sites can be estimated statistically by measuring the total number of phenolic hydroxyls and the methoxyl content of the pyrolysis oil. Analytically the number can be estimated by reacting pyrolysis oils with an aldehyde and then measuring the amount of unreacted aldehyde.

The second important molecular feature is the functionality of the phenolic molecules or the number of reactive sites on a single molecule (f). There are three reactive sites on each phenolic molecule; this number does not vary. But with the pyrolysis oils there is a range of reactive sites. With a number average molecular weight of 400–500 Daltons and a weight average molecular weight of more than 1,000 Daltons *(5)* some phenolics have more than 10 reactive sites on a molecule. As shown in Figure 4b, the number of reactive sites is important since a few molecules with a large number of reactive sites can dominate the gelation behavior. The results in Figures 5a and 5b also suggest that for complex mixtures, which approximate the composition of pyrolysis oils, the functionality of the system plays an important role and may not exactly follow the behavior predicted by equation 2. Although the phenolic mixtures shown in Figures 5a and 5b may show the correct general trends with respect to r and ρ, they do not exhibit the predicted behavior in terms of f. Measuring the average functionality of the system for pyrolysis oils is very difficult, but it can be estimated from the number average molecular weight, the total number of phenolic hydroxyls, and the methoxyl content of the pyrolysis oil.

The third important molecular feature is the number of difunctional and monofunctional phenolic molecules (ρ). Difunctional molecules act as chain extenders; monofunctional phenolic molecules terminate the polymerization reaction. In a crosslinked system, these free chain ends are defects that will reduce the mechanical strength of the network. In these phenolic-rich pyrolysis oils, a significant number of difunctional and monofunctional phenolic compounds are present. The number of these compounds can be measured by gas chromatography, although this number is likely to be low since the pyrolysis oils will contain relatively nonvolatile difunctional and monofunctional phenolic compounds that will be difficult to measure by GC.

Conclusions

This study demonstrated that under the reaction conditions typically used to prepare commercial PF resins, the reaction of phenol (or phenolic model compounds) and formaldehyde follow second-order kinetics. The reaction rate of each compound increases when the reaction temperature and amount of base increases. The reactivity of 2-methoxy-4-methylphenol, 2-methylphenol, or 4-methylcatechol was higher than that of phenol at temperatures and sodium hydroxide concentration used in this study. This indicates that the phenolic compounds commonly found in pyrolysis oils will be highly reactive under traditional PF resin synthesis conditions, and should be chemically incorporated into the PF network.

This work also shows that the complex mixture of phenolics in pyrolysis oils will have well-defined and predictable behavior as they polymerize. As the work of Flory *(23)* and Stockmayer *(24)* predicts, increasing the amount of the phenolic component or the functionality of the individual phenolic molecule, or decreasing the concentration of difunctional and monofunctional phenolic molecules will decrease the reaction time required to reach the gel point.

All these results highlight the need to optimize the PF resin formulation for the pyrolysis oil of interest. The source of the biomass feedstock, the reaction conditions used for the pyrolysis, and the type of pyrolysis reactor can significantly affect the chemical composition of the pyrolysis oil. The results

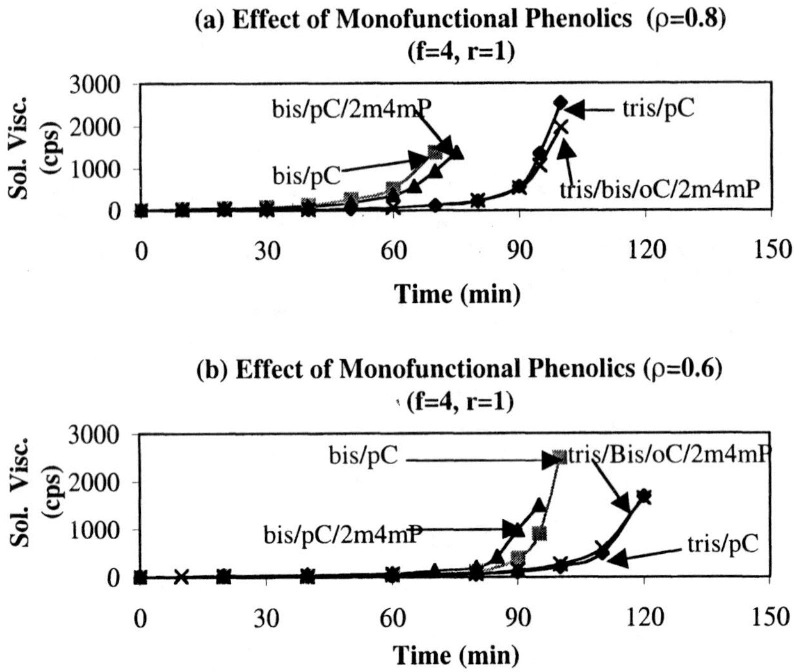

Figure 5. Changes in the solution viscosity as a function of reaction time for complex mixtures of phenolic model compound systems; (a) a series of phenolic model compounds with 20 mole percent of monofunctional phenolic and (b) a series of phenolic model compounds with 40 mole percent of monofunctional phenolic.

presented here can help guide resin chemists as they work to optimize the resin formulation.

Acknowledgments

Support for this work came from the U.S. Department of Energy, Office of Industrial Technologies.

References

1. Hemingway, R. W.; Conner, A. H.; Branham, S. J., Eds. *Adhesives from Renewable Resources*, ACS Symposium Series 385, American Chemical Society, Washington, D.C., 1989.
2. Forss, K.; Fuhrmann, A. *Forest Prod. J.* **1979**, *20* (7): 39–43.
3. Hse, C-Y.; Hong, Q. In *Adhesive from Renewable Resources*, Hemingway, R. W.; Conner, A. H.; Branham, S. J., Eds.; ACS Symposium Series 385, American Chemical Society, Washington, D.C., 1989, pp 97–109.
4. Olivares, M.; Aceituno, H.; Neiman, G.; Rivera, E.; Sellers, T. *Forest Prod. J.* **1995**, *45* (1), 63–67.
5. Chum, H.; Diebold, J; Scahill, J.; Johnson, D.; Black, S.; Schroeder, H.; Kreibich, R. in *Adhesives from Renewable Resources*, Hemingway, R. W.; Conner, A. H.; Branham, S. J., Eds.; ACS Symposium Series 385, American Chemical Society, Washington, D.C., 1989, pp 135–151.
6. Chum, H. L.; Black, S. K. U.S. Patent 4,942,269, 1990.
7. Chum, H. L.; Kreibich, R. E. U.S. Patent 5,091,499, 1992.
8. Chum, H. L.: Black, S. K.; Diebold, J. P.; Kreibich, R. E. U.S. Patent 5,235,021, 1993.
9. Kelley, S. S.; Wang, X-M.; Myers, M. D.; Johnson, D. K.; Scahill, J. W. In *Developments in Thermochemical Biomass Conversion;* Bridgwater, A. V.; Boocock, D. G. B. Eds.; Blackie Academic and Professional, London, 1997, Vol. 1., pp 557–572.
10. Freeman, J. H.; Lewis, C. W. *J. Amer. Chem. Soc.* **1954**, *76,* 2080–2087.
11. Kim, M. G.; Amos, L. W. *Ind. Eng. Chem. Res.* **1990**, *29*, 208–212.
12. Malhotra, H. C.; Gupta, V. K. *J. Appl. Polym. Sci.* **1978**, *22,* 343–351.
13. Malhotra, H. C.; Kumar, V. *J. Macromo. Sci.* **1979**, A*13*, (1), 143–152.
14. Malhotra, H. C.; Tyagi, V. P. *J. Macromol. Sci.* **1980**, A*14,* (5), 675–686
15. Sojka, S. A.; Wolfe, R. A.; Dietz, E. A.; Dannels, B. F. *Macromolecules* **1979**, *12*, 767–770.
16. Sojka, S.A.; Wolfe, R. A.; Guenther, G. D. *Macromolecules* **1981**, *14,* 1539-1543.
17. Pethrich, R.; Thomson, B. *Br. Polym. J.* **1986**, *18* (3), 171–180.
18. Pethrich, R.; Thomson, B. *Br. Polym. J.* **1986**, *18* (6), 380–386.
19. Mukoyama, Y.; Tanno, T.; Yokokawa, H.; Fleming, J. *J. Polym. Sci.* **1973**, *11,* 3193–3204.
20. Bogan, L. E. *Macromolecules* **1991**, *24*, 4807–4812.

21. Carothers, J. A.; Gipstein, E.; Fleming, W. W.; Tompkings, J. A. *J. Appl. Polym. Sci.*, **1982**, *27,* 3449–3451.
22. Fitzgerald, E.A. *J. Appl. Polym. Sci.* **1990**, *41,* 1809–1814.
23. Flory, P. J. *Principles of Polymer Science*; Cornell University Press, NY, 1953; pp 348–378.
24. Stockmayer, W. B. *J. Polym. Sci.* **1952**, *9*, 69–71.
25. Miller, D. R.: Macosko, C. W. *Macromolecules* **1978**, *11,* 656–662.
26. Miller, D. R.: Macosko, C. W. *Macromolecules* **1980**, *12,* 1063–1069.
27. Muller, M.; Burchard, W. *Makromol. Chem.* **1978**, *179,* 1821–1835.
28. Durand, D.; Bruneau, C-M. *J. Polym. Sci., Polym. Phys. Ed.* **1979**, *17,* 273–294.
29. Durand, D.; Bruneau, C-M. *J. Polym. Sci., Polym. Phys. Ed.* **1979**, *17,* 294–303.
30. Maga, J. A. *Food Reviews Internat.* **1987**, *3 (1&2),* 139–183.

Chapter 15

Thermoplastic Polyesters from Steam Exploded Wood

Wolfgang G. Glasser and Rajesh K. Jain

Biobased Materials/Recycling Center, Department of Wood Science and Forest Products, Virginia Polytechnic Institute and State University, Blacksburg, VA 24061

Synopsis

The direct conversion of low-grade woody biomass into useful thermoplastic polymers by chemical modification is illustrated for steam exploded yellow poplar *(Liriodendron tulipifera)* fibers that are esterified in a cellulose-swelling medium. The resulting melt-flowable wood ester mixture was found to be subject to reshaping by melt-processing techniques, or it could optionally be separated into its polymeric constituents (cellulose ester and lignin ester) by choice of appropriate nonsolvent during recovery. The reaction conditions provide for the generation of a thermoplastic cellulose ester/lignin ester mixture with single glass transition temperature that was found to vary with both degree of substitution and lignin content. The esterification and optional fractionation of steam exploded wood is perceived as a possibly simple and inexpensive route to thermoplastic structural polymers from wood.

[1] This represents Part 6 of a series of articles on steam assisted fractionation of biomass. Earlier papers in this series have appeared in Biomass and Bioenergy, Bioresource Technology, and the Journal of Applied Polymer Science.

Introduction

The vast majority of renewable natural resources comprises materials that consist of several polymeric constituents *(1-3)*. Wood is a multicomponent material composed of principally three polymers, a crystalline homopolysaccharide (cellulose), several types of branched, non-ordered heteropolysaccharides (hemicelluloses), and an apparently non-ordered polyaromatic polymer that shows network-like behavior (lignin) *(4)*. While the chemistry of all three constituents is by and large well understood, the basis for the interaction of these three polymers is still subject to some conjecture *(5,6)*.

The separation of constitutive polymers from wood and related woody biomass is the subject of dozens of commercial and semicommercial separation processes, including, but not limited to, pulping processes *(7,8)*. In addition, there are hundreds of patents describing technology for the isolation of constitutive plant polymers. Commercially successful wood fractionation technologies usually focus on the separation of a single end product, with the rest being discarded or incinerated. This is the case for pulping processes as well as for wood saccharification and furfural production. It is apparently easier to optimize a separation technology for the purity and performance requirements of a single end product without having to be concerned with the remainder. Although there have been serious attempts at multiproduct processes, such as those aiming at the isolation of cellulose and lignin by organosolv pulping *(9,10)*, these processes have largely failed, in part because of a mismatching of market size for the individual end products.

In addition to acquiring knowledge about the chemical composition of all wood-based polymers (cellulose, hemicelluloses, and lignin *[11]*), much has been learned about controlling product characteristics and properties via chemical modification *(12-14)*. In addition, it has been demonstrated that modifications can often and conveniently become an integral part of the isolation protocol *(15)*. Polymer modification for the purpose of achieving desired end product properties represents a long-accepted commercial practice with naturally-occurring polymers *(16,17)*. Examples are cellulose esters which become soluble and/or thermoplastic in relation to substituent type and degree of substitution *(18)*; thermoplastic starch esters that are capturing markets for a variety of biodegradable materials *(19)*; lignin products whose reactivity, solubility, and compatibility with other polymer systems depends on source, size, purity, etc. *(20-23)*; and xylan ether products that become water soluble and/or thermoplastic in relation to chemical modification *(15)*.

The conversion of natural resources into a macromolecular products with useful properties usually constitutes the process step that follows the isolation of the polymer in question. However, the sequence of separation-isolation-modification, which is so widely practiced throughout the industry dealing with naturally-occurring polymers, is not always a requirement inherent to the polymer in question. It is conceivable that separation and isolation follow modification in those cases in which the presence of one component does not interfere with the modification of the other, or where both (or all) components in the mixture are modified together, only to be separated following the modification.

The chemical modification of unmodified wood with the consequence of the formation of thermoplastic or even liquid wood derivatives has been the subject of work by Shiraishi et al. *(24,25)*. This approach is constrained by the strong interaction between lignin and hemicelluloses which usually prevents the formation of a derivative that is soluble in a common solvent, or that is melt flowable. However, after the mild hydrolysis of hemicelluloses and of lignin-carbohydrate bonds, by a procedure such as steam explosion *(26)*, the cellulose-lignin mixture has been found to provide increased accessibility to solvents, enzymes, and derivatizing agents *(27)*, and to produce homogeneous solutions in appropriate solvents from which the mixture can be regenerated *(28,29)*. Following (partial) hydrolysis of the hemicelluloses, lignin gains the freedom to undergo flow at temperatures above its glass transition temperature *(30)*. This mobility is responsible for the ease of fractionation using the steam explosion technique *(31)*.

It was the objective of this study to jointly modify lignin and cellulose in steam exploded wood using acid anhydrides for modification. Molecular mixtures of cellulose esters and lignin esters have recently been investigated by Ghosh et al. *(32)* in terms of compatibility features. Although phase separation had been demonstrated for virtually all blends, single glass transitions were reported for some blends that followed the Fox equation. The goals of this study therefore dealt with generating cellulose ester/lignin ester blends directly from steam exploded wood. Such thermoplastic polyester mixtures can be expected to find use in melt-processed (plastic) components for the composites, furniture, automotive, etc., industries. The study was also to examine avenues for the efficient separation of the individual polymer components of the ester blend.

Cellulose esterification with acetic, propionic, and butyric anhydride is a widely practiced industrial technology that results in solvent-soluble and thermoplastic cellulose ester derivatives *(18)*. A new esterification method has recently been introduced by Edgar et al. *(33)* which relies (a) on the reaction of cellulose in a solvent medium capable of swelling cellulose, and (b) on using the anhydrides of long chain fatty acids in combination with a catalyst (titanium isopropoxide). This method has been shown to produce cellulose esters with degrees of substitution between 2.4 and 2.8 using stoichiometric quantities of anhydrides. The resulting ester products were found to have melt-rheological properties that are normally found only in plasticized esters *(33)*.

It was the objective of the current study to examine the application of this modification technique to steam exploded (water-washed) wood consisting essentially of a physical mixture of lignin and cellulose.

Materials and Methods

1. *Steam Explosion*: Steam explosion was conducted in a two-cubic foot batch digester equipped with steam injection and ball valves providing for rapid decompression. Approximately 2 kg of yellow poplar (*Liriodendron tulipifera*) wood chips were loaded batchwise into the vertical reaction vessel and treated with live steam (232°C) for 2.5 minutes. Using the method of Overend and Chornet (34), these conditions corresponded to a severity of log R_0 4.3. Following decompression, the exploded fibrous mulch was collected from a cyclone steam separator.

2. *Water Washing*: The fibrous mulch was extracted with hot water in accordance with the protocol given earlier *(31,35)*. Fiber solids were recovered by centrifugation and washing. Fiber solids content was typically 30%.

3. *Organosolv Extraction*: The water washed fibers were subsequently extracted with hot glacial acetic acid (60°C) for one hour. Acetic acid-solubles were separated from insoluble fiber solids by centrifugation and washing with acetic acid.

4. *Acylation* (Hexanoylation): The acetic acid-containing (i.e., solvent-exchanged) cellulose-rich fibers (solids content, ca. 30%; lignin content, ca. 12%) were suspended in dimethyl acetamide (DMAc) in a large volume round bottom flask attached to a reflux condenser. The mixture was charged with 2 moles of hexanoic anhydride per number of hydroxy equivalents in the fiber solids. The OH-equivalent content of the fiber solids was assumed to be identical to cellulose (i.e., 18.5 meq/g). To this mixture was added 0.2% by weight titanium isopropoxide on fiber solids. The mixture was refluxed for between 4 and 24 hours before it was cooled and precipitated into cold water or isopropanol. The respective precipitants were recovered by filtration and drying.

5. *Analysis*: Molecular weight determinations were performed using the tricarbanilate procedure according to Wallis et al. *(36)*, on a high pressure liquid chromatography system equipped with refractive index detector and differential viscometer. THF served as mobile phase, and a series of three micro-Spherogel columns (Waters Assoc.) having exclusion volumes of 100, 1000, and 10,000 A were used for separation. Thermal analysis for glass transition temperatures involved a DSC-4 differential scanning calorimeter by Perkin and Elmer (System 4) and a dynamic mechanical thermal analyzer (DMTA) (Polymer Laboratories, Ltd.). The degree of substitution was determined using a modified FTIR-method according to Glasser et al. *(37,38)* based on a commercial cellulose acetate sample (DS 2.45) as standard. Lignin content was determined

by UV-spectroscopy in acetone solution at 280 nm in accordance with standard protocol *(39)*. Isolated lignin ester was used for standard.

Results and Discussion

The fractionation of steam exploded woody biomass into constitutive molecular components is usually accomplished using the processing scheme illustrated in Figure 1 (adapted to organosolv delignification with acetic acid) *(31)*. Moist steam exploded fiber is thereby washed with hot water, followed by extraction with warm acetic acid in place of the more typical aqueous alkali. The acetic acid-soluble fraction contains mostly lignin which can be separated from the extraction mixture by evaporation/precipitation. Conversely, the unbleached solid fiber residue can be isolated as a 75-90% cellulose-containing solid by washing and drying. Steam exploded fibers are more commonly extracted with alkali following water washing with the consequence of the removal of alkali-soluble lignin and the generation of a carbohydrate-rich fiber fraction *(31)*. Both fractionation protocols have their respective advantages and disadvantages that are related to solvent versus alkali recovery, in principle. Alkali-extraction is usually more efficient for delignification *(35)*.

In the case in which acetic acid is used as the extracting solvent according to Figure 1, an acetic acid-containing cellulose fiber emerges which has undergone substantial solvent exchange from water to acetic acid. If this solvent-exchanged fiber is suspended in a cellulose-swelling solvent medium, like dimethyl acetamide (DMAc), the fiber can be subjected to an esterification reaction in accordance with the protocol of Edgar et al. *(33)*. The advantages of using titanium-isopropoxide (Ti-IP) as catalyst have been described as (a) avoidance of peracylation followed by partial hydrolytic deacylation, and (b) ability to modify with larger acyl substituents than butyrate *(33)*. Following exposure to a nearly stoichiometric amount of anhydride, the cellulose/lignin fiber dissolves in the solvent mixture, which can be fully recovered. This dissolution must be considered the critical aspect of this procedure. Whereas no dissolution would occur with wood (sawdust) as starting material, virtually complete dissolution is achieved with steam exploded fibrous mulch after 12 hr (Figure 2). Degree of substitution (DS) thereby is related to reagent stoichiometry and reaction time (Figure 3).

Following precipitation of the reaction mixture into water, the dissolved wood esters can be recovered in powder form with a color virtually identical to the original steam exploded and water-washed fibrous mulch. The catalyst thereby co-precipitates forming an ash (TiO_2)-containing polyester mixture.

The thermal properties of the regenerated wood ester product reveal a single glass transition temperature (T_g) (Fig. 4) which varies with both degree of substitution (DS) (i.e., reaction-time) (Fig. 5) and (to a lesser extent) lignin

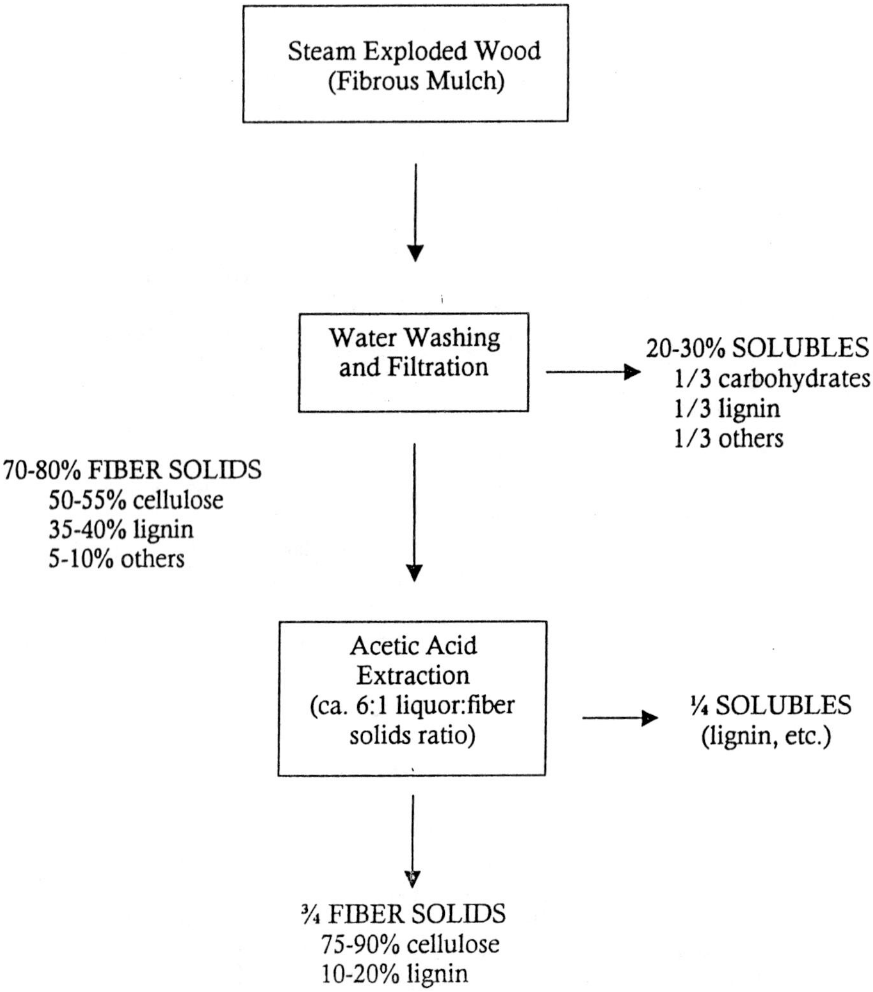

Fig. 1. Typical Process Scheme for the Fractionation of Steam Exploded Wood Using Organosolv Delignification with Acetic Acid.

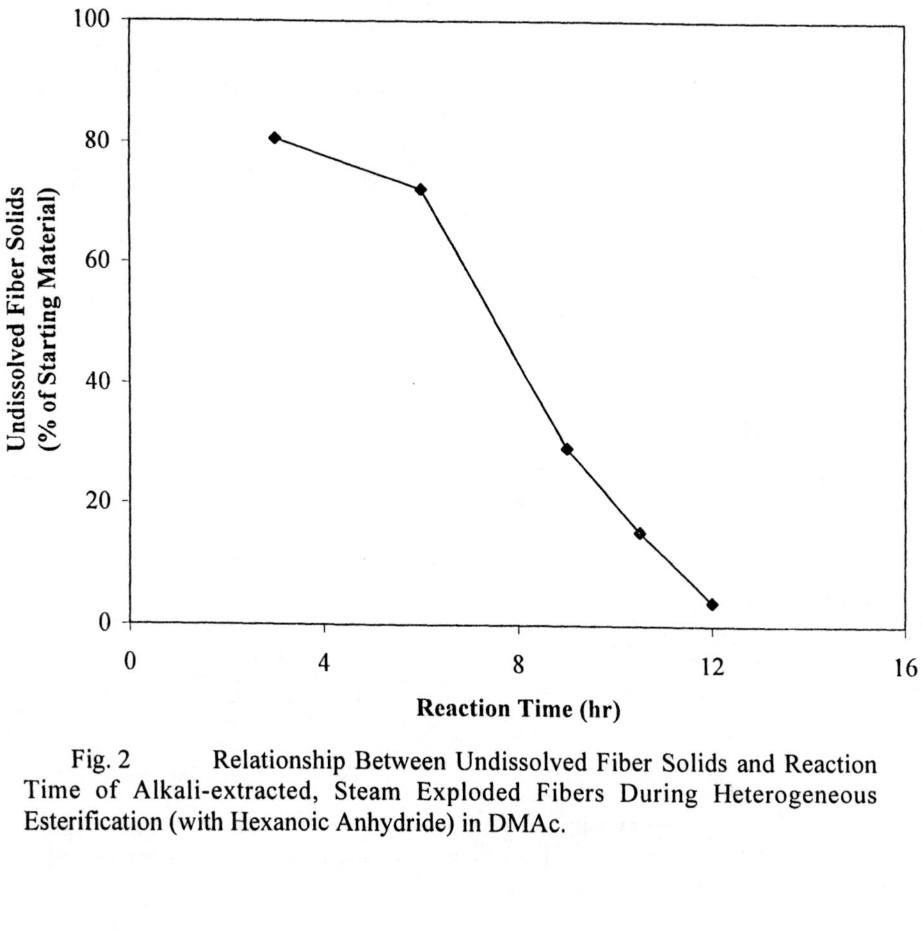

Fig. 2 Relationship Between Undissolved Fiber Solids and Reaction Time of Alkali-extracted, Steam Exploded Fibers During Heterogeneous Esterification (with Hexanoic Anhydride) in DMAc.

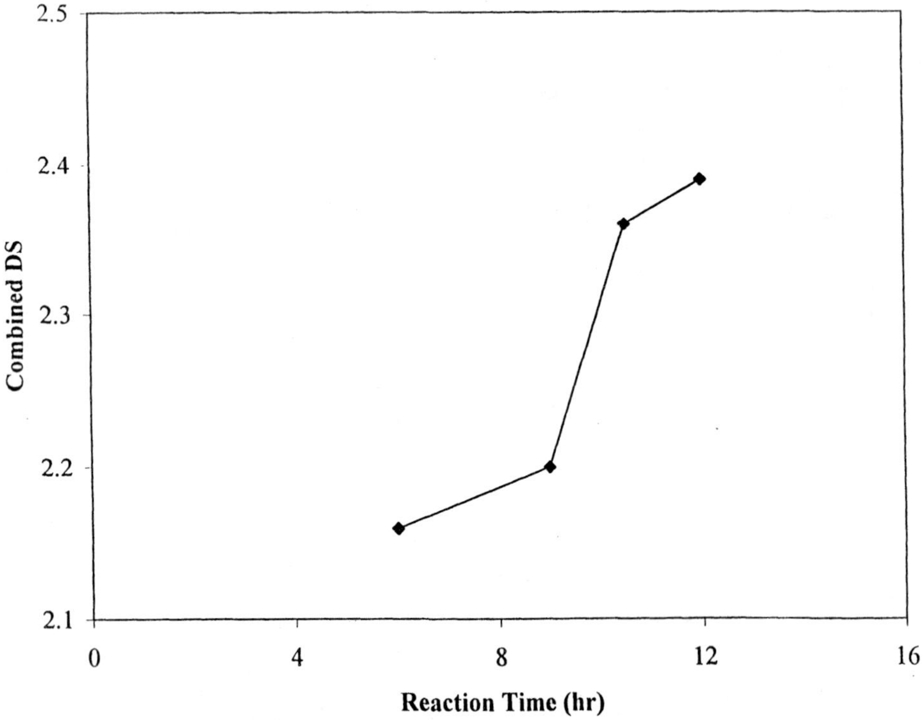

Fig. 3 Relationship between (combined) DS and Reaction Time of Alkali-extracted, Steam Exploded Fibers During Heterogeneous Esterification (with Hexanoic Anhydride) in DMAc.

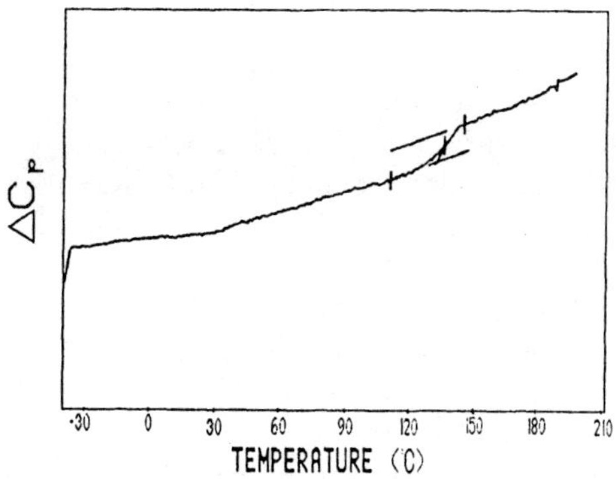

Fig. 4 DSC-Thermogram of Peracetylated Polyester Mixture from Water-Precipitated, Hexanoylated Steam Exploded Fibers. Lignin had been extracted with alkali.

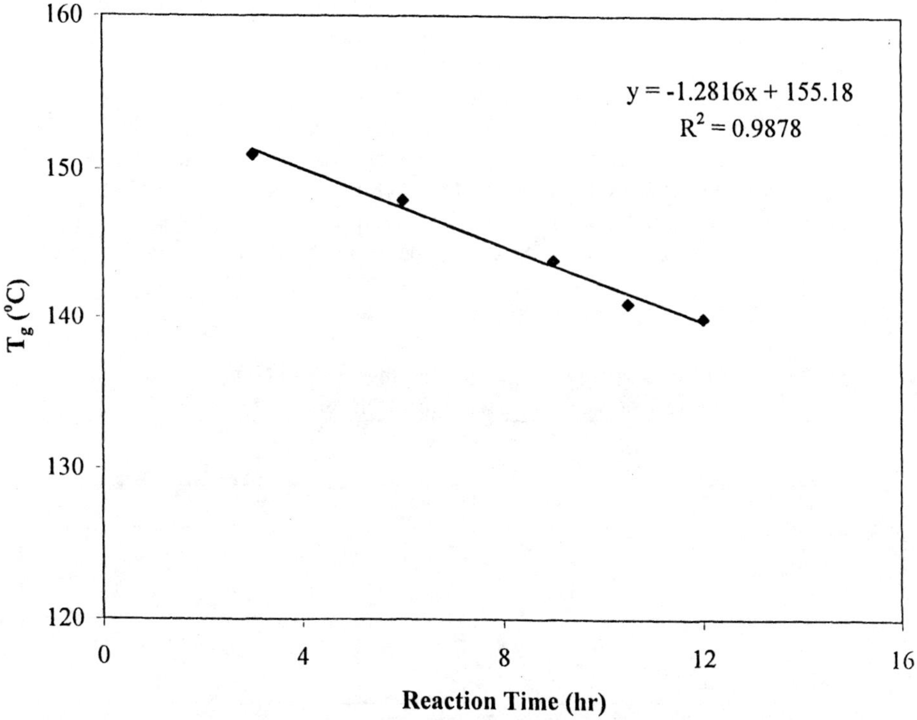

Fig. 5 Relationship Between Reaction Time and Glass Transition Temperature (T_g) of Polyester Mixture from Water-Precipitated, Hexanoylated Steam Exploded Fibers.

content (Figure 6). A single T_g thereby implies either copolymer structure with absence of significant phase separation or the existence of a compatible or miscible molecular mixture of cellulose esters and lignin esters *(32,40)*. Analysis by gel permeation chromatography (GPC) reveals the existence of two distinct components (Figure 7) which appear as a minor, low molecular weight component (lignin ester) and a major, high molecular weight constituent (cellulose ester). The molecular weight of the cellulose ester component declines with reaction time (Table I) revealing (again) the competitive nature of depolymerization/derivatization *(33)*.

The use of an organic solvent which is qualified to dissolve lignin esters, but which is unqualified to solvate cellulose esters, for precipitation of the reaction mixture allows for the separation of the two components into individual polymeric constituents with virtually no cross contamination (Figure 8).

This observation, of the presence of lignin and cellulose in physically mixed rather than copolymeric form, is in agreement with observations by Westermark et al. who found that unbleached kraft pulp fibers from birch can be dissolved in DMAc/LiCl with an apparent separation of lignin-rich and cellulose-rich molecular constituents by gel permeation chromatography *(41)*.

This suggests that cellulose esters may be manufactured following a simplified esterification protocol in which steam exploded fibers are subjected to chemical modification followed by an optional separation of lignin at the end of the esterification procedure (Figure 8). The protocol is capable of generating an impure wood ester product with engineerable thermoplastic properties, or it may conversely be used to prepare a pure cellulose ester product, cleanly separated from a lignin ester polymer by appropriate solvent precipitation.

Table I. Number-average Degree of Polymerization (DP_n) of Cellulose Ester Derivatives in Relation to Reaction Time

Reaction time (hr.)	DP_n [1)]
9	2650
10.5	1420
12	530

[1)] By GPC in THF using cellulose tricarbanilate derivatives

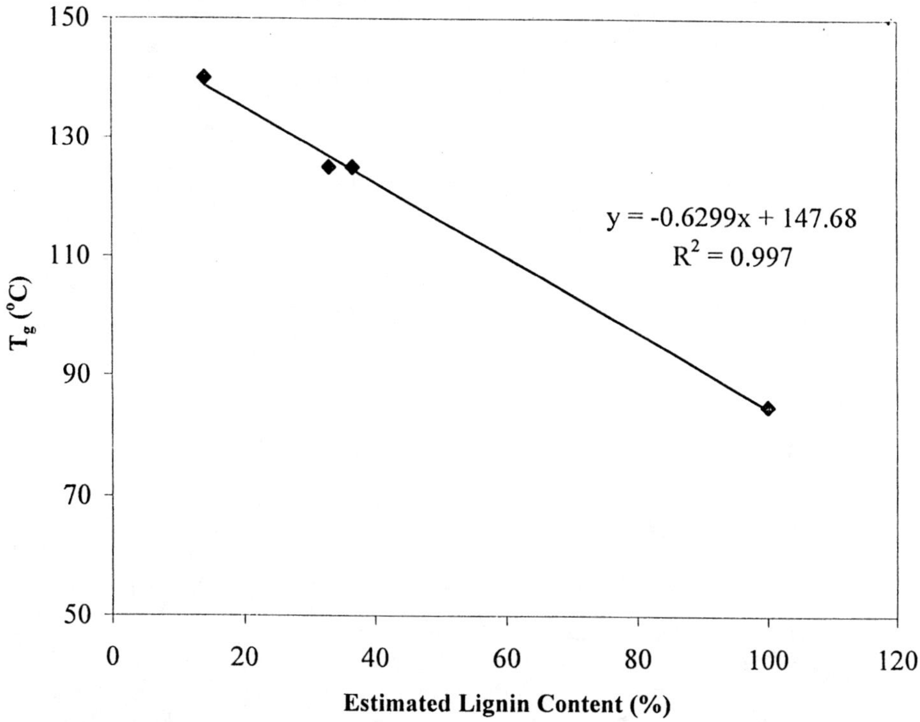

Fig. 6 Relationship Between T_g and Estimated (by UV) Lignin Content of Polyester Mixture from Hexanoylated Steam Exploded Fibers. Lignin had been extracted with alkali in case of the sample with the lowest lignin content; and it had been extracted with acetic acid in case of the two fiber samples with the higher lignin content.

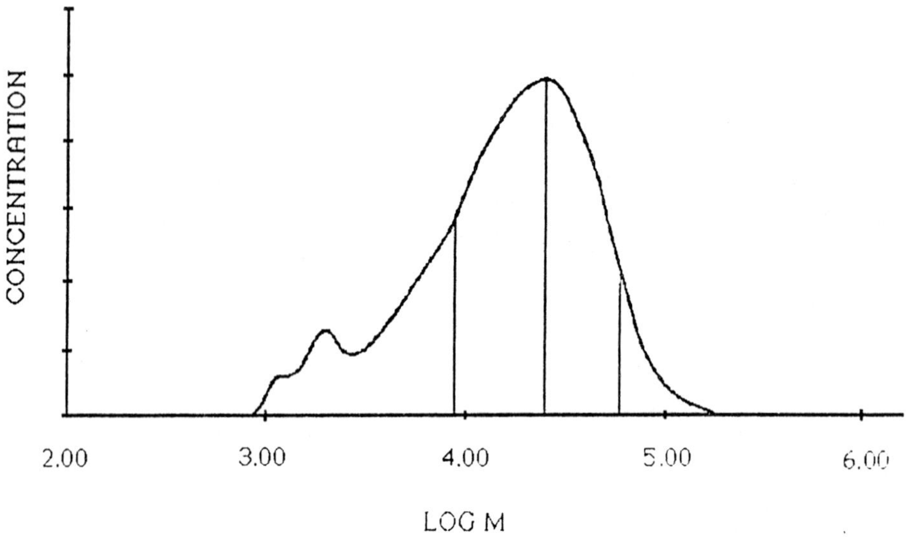

Fig. 7 Gel Permeation Chromatogram of Hexanoylated, Steam Exploded Fiber Solids Precipitated in Water. A corresponding sample precipitated in isopropanol (good solvent for lignin ester) revealed an almost uniform distribution curve from which the low molecular weight constituents (i.e., lignin ester) were missing.

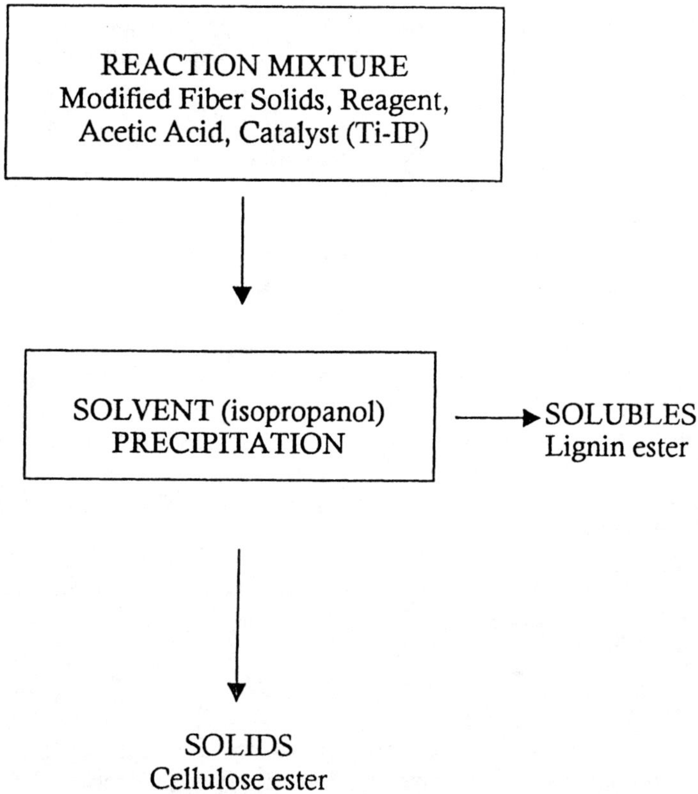

Fig. 8 Flow Diagram for the Separation of Cellulose Esters from Lignin Esters Using Esterified Steam Exploded Fibers.

The separation of lignin esters from cellulose esters for use in thermoplastics may not always be necessary. In circumstances where color and clarity are unessential, and where high modulus is desirable, thermoplastic wood esters with significant phase compatibility may be employed as virtually uniform polymeric material *(32,40)*. The positive interaction between lignin and cellulose, which is surprising in the face of the recognized substantial chemical differences, has been pointed out in previous research (*5* and references cited therein).

Conclusions

Useful polymers on the basis of natural resources may not only emerge from the polymerization of low molecular weight chemicals, obtained by the destructive depolymerization of woody biomass, but they can also be prepared by the direct modification of indigenous polymers contained in a low-grade fibrous mulch obtained by steam explosion.

Properties of naturally occurring polymers can, in general, be controlled by known modification reactions.

Chemical modification can become a part of the overall isolation/separation protocol. Steam exploded biomass can be esterified heterogeneously to produce uniform regenerated wood esters or individual (separated) cellulose ester and lignin ester polymers.

The process sequence of steam explosion-esterification offers a novel and inexpensive route to the production of thermoplastic polyesters from wood.

Acknowledgement

The authors wish to express their gratitude for technical assistance to Mr. Robert S. Wright, Department of Wood Science & Forest Products.

References

1. Society for Experimental Biology. Symposium no. 34, *The Mechanical Properties of Biological Materials*; Cambridge University Press, Cambridge, UK, 1980, 513 pp.
2. Vincent, J. *Structural Biomaterials*; 2nd ed., Princeton University Press, Princeton, NJ, 1991, 244 pp.
3. National Research Council, *Hierarchical Structures in Biology As a Guide for New Materials Technology*; National Academy Press, Washington, DC, 1994, 130 pp.

4. Jung, H. G.; Bruxton, D. R.; Hatfield, R. D.; Ralph, J. *Forage Cell Wall Structure and Digestibility*; American Society of Agronomy, Inc., Madison, WI, 1993, 794 pp.
5. Glasser, W. G.; Rials, T. G.; Kelley, S. S.; Dave, V. Chapter 19 in *Cellulose Derivatives – Modification, Characterization, and Nanostructures*, Heinze, Th. J.; Glasser, W. G. eds., American Chemical Society, Washington, D. C., ACS Symp. Ser. 688, 265-282 (1998).
6. Terashima, N.; K. Fukushima, L.-F. He; Takabe, K. In ref. 4, p. 247-270.
7. Young, R. A.; Masood, A. eds., *Environmentally Friendly Technologies for the Pulp and Paper Industry*, John Wiley & Sons, Inc., New York, 1998.
8. Focher, B.; Marzetti, A.; Crescenzi, V., eds., *Steam Explosion Techniques – Fundamentals and Industrial Applications*; Gordon and Breach Sci. Publishers, Philadelphia, PA, 1988, 413 pp.
9. Hergert, H. L. In. ref. 7, p. 8-69.
10. Lora, J. H.; Aziz, S. *TAPPI* **1985**, 68(8), 94.
11. Fengel, D.; Wegener, G. *Wood-Chemistry, Ultrastructure, Reactions*; Walter de Gruyter, Berlin, 1984, 613 pp.
12. Hon, D. N.-S.; Shiraishi, N., eds. *Wood and Cellulosic Chemistry*; Marcel Dekker, Inc., New York, NY, 1990, 1020 pg.
13. Shiraishi, N. Chapter 18 in ref. 12, p. 861-906 and references cited therein.
14. Rowell, R. M. Chapter 11 in *Paper and Composites from Agro-Based Resources*; Rowell, R. M.; Young, R. A.; Rowell, J. K. eds., CRC Lewis Publishers, Boca Raton, FL, 1997, 351-376 and references cited therein.
15. Glasser, W. G.; Jain, R. K.; Sjostedt, M. A. U.S. Pat. # 5,523,398 (1995).
16. Ishizu, A. Chapter 16 in ref. 12, p. 757-800.
17. Glass, J. E.; Swift, G., eds. *Agricultural and Synthetic Polymers-Biodegradability and Utilization*; American Chemical Society, Washington, D.C., 1990, ACS Symp. Ser. No. 433, 323 pp.
18. Bogan, R. T.; Brewer, R. J., Enc. Polym. Sci. Eng., Vol. 3, 1987, p. 181-226.
19. Narayan, R. In *Assessment of Biobased Materials* H. L. Chum, ed., Solar Energy Research Institute, Colorado SERI/TR-234-3610, 1989, pg. 7.1-7.25.
20. Kelley, S. S.; Glasser, W. G.; Ward, T. C. *J. Wood Chem. Technol.* **1988**, *8*, 341-359.
21. Glasser, W. G.; Jain, R. K. *Holzforschung* **1993**, *47*, 225-233.
22. De Oliveira, W.; Glasser, W. G. *Macromolecules* **1994**, *27*, 5-11.
23. De Oliveira, W.; Glasser, W. G. *J. Appl. Polym. Sci.* **1994**, *51*, 563-571.
24. Shiraishi, N. In ref. 12, pg. 861-906.
25. Yoshioka, M.; Haraguchi, N.; Shiraishi, N. *Cellulose* **1999**, *6*,193-212.
26. Focher, B.; Marzetti, A.; Crescenzi, V. eds. *Steam Explosion Techniques*; Gordon and Breach Sci. Publ., Philadelphia, PA, 1991, 413 pp.
27. Marchessault, R. H., Chapter in ref. 26, 1-19.

28. Chanzy, H.; Parillet, P.; Peguy, A. *Polymer Commun.* **1986,** *27*, 171.
29. Samaranayake, G.; Li, X.; Glasser, W. G. *Holzforschung* **1994**, *48* (Supplement), 69-71.
30. Marchessault, R. H.; St. Pierre, J. In *Future Sources of Organic Raw Materials*; St. Pierre, L.; Brown, G. R. eds., Pergammon Press, Elmsford, NY, 1980, p. 613-626.
31. Wright, R. S.; Glasser, W. G. *Biomass Bioenergy* **1998**, *14*, 219-235.
32. Ghosh, I.; Jain, R. K.; Glasser, W. G. *J. Appl. Polym. Sci.* **1999**, *74*, 448-457.
33. Edgar, K. J.; Pecorini, T. J.; Glasser, W. G. Chapter 3 in *Cellulose Derivatives—Modification, Characterization, and Nanostructures*; Heinze, T. J.; Glasser, W. G. eds., American Chemical Society, Washington, D.C., ACS Symp. Ser. No. 688, 1998, p. 38-60.
34. Overend, R. P.; Chornet, E. *Phil. Transactions of the Royal Society, London A* **1987**, *321*, 523-536.
35. Ibrahim, M.; Glasser, W. G. *Bioresource Technol.* **1999**, *70*, 181-192.
36. Evans, R.; Wearne, R. H.; Wallis, A. F. A. *J. Appl. Polym. Sci.* **1989**, *37*, 3291.
37. Glasser, W. G.; Jain, R. K. *Holzforschung* **1993**, *47*, 225-233.
38. Dhara, K.; Jain, R. K.; Glasser, W. G. *Holzforschung* **1993**, *47*, 403-411.
39. Iiyama, K.; Wallis, A. F. A. *Wood Sci. Technol.* **1998**, *22*, 271-280.
40. Ghosh, I.; Jain, R. K.; Glasser, W. G. Chapter 17 in *Lignin: Historical, Biological, and Materials Perspectives*; Glasser, W. G.; Northey, R. A.; Schultz, T. P. eds., American Chemical Society, Washington, D.C., 1999, ACS Symp. Ser. No. 742, p. 331-350.
41. Karlsson, O.; Westermark, U. *J. Pulp Pap. Sci.* **1996**, *22*, J397-401.

Author Index

Subject Index

B

C

D

H

I

L

M

N

O

P

Q

R

S

T

U

V

W

X

Bestsellers from ACS Books

The ACS Style Guide: A Manual for Authors and Editors (2nd Edition)
Edited by Janet S. Dodd
470 pp; clothbound ISBN 0–8412–3461–2; paperback ISBN 0–8412–3462–0

Writing the Laboratory Notebook
By Howard M. Kanare
145 pp; clothbound ISBN 0–8412–0906–5; paperback ISBN 0–8412–0933–2

Career Transitions for Chemists
By Dorothy P. Rodmann, Donald D. Bly, Frederick H. Owens, and Anne-Claire Anderson
240 pp; clothbound ISBN 0–8412–3052–8; paperback ISBN 0–8412–3038–2

Chemical Activities (student and teacher editions)
By Christie L. Borgford and Lee R. Summerlin
330 pp; spiralbound ISBN 0–8412–1417–4; teacher edition, ISBN 0–8412–1416–6

Chemical Demonstrations: A Sourcebook for Teachers, Volumes 1 and 2, Second Edition
Volume 1 by Lee R. Summerlin and James L. Ealy, Jr.
198 pp; spiralbound ISBN 0–8412–1481–6
Volume 2 by Lee R. Summerlin, Christie L. Borgford, and Julie B. Ealy
234 pp; spiralbound ISBN 0–8412–1535–9

The Internet: A Guide for Chemists
Edited by Steven M. Bachrach
360 pp; clothbound ISBN 0–8412–3223–7; paperback ISBN 0–8412–3224–5

Laboratory Waste Management: A Guidebook
ACS Task Force on Laboratory Waste Management
250 pp; clothbound ISBN 0–8412–2735–7; paperback ISBN 0–8412–2849–3

Reagent Chemicals, Ninth Edition
768 pp; clothbound ISBN 0–8412–3671–2

Good Laboratory Practice Standards: Applications for Field and Laboratory Studies
Edited by Willa Y. Garner, Maureen S. Barge, and James P. Ussary
571 pp; clothbound ISBN 0–8412–2192–8

For further information contact:
Order Department
Oxford University Press
2001 Evans Road
Cary, NC 27513
Phone: 1-800-445-9714 or 919-677-0977